EVIDENCE

Eight distinguished experts from a wide range of disciplines consider the nature and use of evidence in the modern world. Peter Lipton begins the book by analysing evidence in general philosophical terms. Carlo Ginzburg then examines the ambiguities of historical evidence. Vincent Courtillot analyses the evidence for cataclysmic geological change. Monica M. Grady considers the evidence for life in space. Brian Greene discusses superstring theory and the quest for a unified theory of the Universe. Philip Dawid explores the uses and abuses of statistical evidence in landmark legal cases, while Cherie Booth looks at the role of evidence in domestic and international law. In the final chapter Karen Armstrong considers the role of evidence in religious belief.

THE DARWIN COLLEGE LECTURES

EVIDENCE

Edited by
Andrew Bell, John Swenson-Wright and Karin Tybjerg

CAMBRIDGE
UNIVERSITY PRESS

CAMBRIDGE UNIVERSITY PRESS
Cambridge, New York, Melbourne, Madrid, Cape Town, Singapore, São Paulo, Delhi

Cambridge University Press
The Edinburgh Building, Cambridge CB2 8RU, UK

Published in the United States of America by Cambridge University Press, New York

www.cambridge.org
Information on this title: www.cambridge.org/9780521710190

First published 2008

A catalogue record for this publication is available from the British Library

The quotations on pages 1 and 2 are taken from *Winnie-the Pooh,* by A.A. Milne,
first published in 1926 by Methuen and Co. Ltd.

ISBN 978-0-521-83961-7 hardback
ISBN 978-0-521-71019-0 paperback

Transferred to digital printing 2009

In memory of Peter Lipton

Contents

Introduction

ANDREW BELL, JOHN SWENSON-WRIGHT AND KARIN TYBJERG

Evidence is central to how we understand the world and how we go about persuading others that we are right. Sometimes its identification and interpretation are straightforward. Daniel Defoe's *Robinson Crusoe* presents a particularly clear example of the power of evidence. Believing himself alone on an island, Crusoe one day discovers a footstep in the sand:

> It happen'd one day about noon going towards my boat, I was exceedingly surpris'd with the print of a man's naked foot on the shore, which was very plain to be seen in the sand: I stood like one thunder-struck, or if I had seen an apparition . . . after innumerable fluttering thoughts like a man perfectly confus'd and out of myself, I came home to my fortification, not feeling, as we say, the ground I went on, but terrify'd to the last degree, looking behind me at every two or three steps, mistaking every bush, and fancying every stump at a distance to be a man.

Crusoe reacts with shock and surprise to this unsettling evidence. It seems to have only one possible interpretation; we do not even need to be told that Crusoe now knows that he is not alone. His conclusion is inescapable and this one piece of evidence changes everything about his situation in a stroke. For those concerned with what evidence is and how it works, this is an ideal-case scenario. The footprint establishes new knowledge in an instant with little room for argument. We often find ourselves hoping that the evidence we put forward for our own beliefs is equally compelling.

Unfortunately, evidence is not always evident and does not always present itself to us in so readily interpretable a fashion. Winnie-the-Pooh and Piglet find their case of footprints more complex:

> 'Hallo!' said Piglet, 'what are *you* doing?' . . .
> 'Tracking something,' said Winnie-the-Pooh very mysteriously.
> 'Tracking what?' said Piglet coming closer.

> 'That's just what I ask myself. I ask myself, What?' . . .
>
> 'Now look there.' He pointed at the ground in front of him 'What do you see there?'
>
> 'Tracks,' said Piglet. 'Paw-marks.' He gave a little squeak of excitement. 'Oh, Pooh! Do you think it's a – a – a Woozle?'
>
> 'It may be,' said Pooh. 'Sometimes it is, and sometimes it isn't. You never can tell with paw-marks.'
>
> With these few words he stopped tracking, and Piglet, after watching him for a minute or two, ran after him. Winnie-the-pooh had come to a sudden stop, and was bending over the tracks in a puzzled sort of way.
>
> 'What's the matter?' asked Piglet.
>
> 'It's a funny thing,' said Bear, 'but there seem to be *two* animals now. This – whatever-it-was – has been joined by another – whatever-it-is – and the two of them are now proceeding in company. Would you mind coming with me, in case they turn out to be Hostile Animals?'

Pooh and Piglet do not know what they are looking for and they do not even know if they really want to find it. The plot thickens as the tracks multiply with more Woozles and Wizzles seemingly joining the fray. But with a little help Pooh suddenly realizes the provenance of the tracks:

> Then he fitted his paw into one of the Tracks . . . and then he scratched his nose twice and stood up.
>
> 'Yes,' said Winnie the Pooh.
>
> 'I see now,' said Winnie the Pooh.
>
> 'I have been Foolish and Deluded,' said he, 'and I am a Bear of No Brain at All.'

But Winnie-the-Pooh is no more a Bear of No Brain at All than many distinguished minds searching for knowledge. Separating our own tracks from those we follow in our quest for knowledge is not easy. As soon as we involve our own reasoning, our own assumptions, in the search for evidence, we leave our own footprints. Often we allow these footprints, in the guise of intuition or experience, to guide us. Furthermore, recognizing those tracks which lead to particular answers among the scores of others is no easier. We do not choose to follow all the tracks in the world, indeed we choose actively to ignore most. When we hunt for knowledge we almost always hunt with an idea of what we want to find.

In bringing together authors from a wide range of disciplines, this collection, based on the nineteenth Darwin public lecture series, seeks to show how different practices and assumptions lie behind claims to know. Evidence comes in a multiplicity of forms and each discipline or problem requires its own standards and kinds of evidence: witnesses, statistics, documents, experiments, mathematical consistency, norms or rules of reasoning, even divine revelation. The way in which evidence is used, accepted and challenged varies widely. At times, it can seem as though different practitioners are not even talking about the same thing. For many, the use of evidence is simply a matter of following good practice in their chosen field, be it detective work or theoretical physics. But the business of evidence plays a central role in all generation of knowledge. Exploring the nature and use of evidence *across* different disciplines gives some insight into what it means to claim to know or to dispute knowledge. The authors in this volume focus on controversial questions and it is in the midst of controversy that problems of evidence become most apparent.

The problem of evidence is often thought to be the province of philosophers who discuss questions of knowledge in general terms and thus highlight the structures of inference that underlie both everyday and scientific thinking. In studying how knowledge is generated, justified and mediated in particular cases, we can gain a still richer understanding. The following essays offer both a philosophical account and case studies of how evidence is used in different disciplines, from geology to law, physics to religion. We start with history, which rarely provides us with clear tracks to pursue, as often the materials available for study are themselves heavily overlain with the traces of their own circumstances and influences. Turning to the sciences, it becomes clear that there is no one scientific method for dealing with evidence. There is a multiplicity of approaches and often priorities must be made about which tracks should be followed in order to answer complex questions. Finally, three contributors consider how the needs of society and the practice of religion place other demands on the ways we approach evidence.

Peter Lipton brings what he calls 'the characteristically abstract perspective of the philosopher' to the problem of what evidence is and how we use evidence to generate knowledge about the world. Our beliefs about the world – namely, the many facts we gather through perception and testimony from others – can be seen as a vast network. We can expand this network both by learning new things about the world, and by combining what we know – drawing inferences – to form new beliefs. The beliefs we use to create new beliefs within the network serve as evidence and are essential to extending our knowledge. For instance, the arrival of letters is evidence that the postman has called. A famous problem facing us when we draw inferences that go beyond logical consequences is that they are based on an assumption that things in the future will be as they are now. We expect more of the same! But we can never be absolutely certain that this is the case. No matter how many times the Sun rises there is no guarantee that it will do so tomorrow. We do, however, routinely use this form of reasoning and we even feel justified in doing so. Most contributors to this volume work on the assumption that 'more of the same' is a reasonable basis on which to draw inferences. Lipton finds the answer to why we rely on such inferences in our love of explanations. In this way he turns the problem on its head. We do not simply generate explanations on the basis of evidence. Certain beliefs become important to us – become evidence – exactly because they generate an elegant and satisfying explanation.

Moving beyond Lipton's philosophical enquiry, Carlo Ginzburg illustrates, in the first case study, the difficulties and dangers of the multiple tracks found in historical enquiry. At the heart of his essay is the assumption that all historical evidence is to some degree ambiguous and that its ambiguity can be limited by carefully establishing context. Ginzburg tackles two difficult pieces: a fictional text and a forgery. The fictional text is a mid nineteenth-century book, *A Dialogue in Hell between Machiavelli and Montesquieu* by Maurice Joly. In this text, the ghosts of these two thinkers debate methods for the political manipulation of society. The forgery is the infamous *Protocols of the Sages of Zion*, an anti-Semitic pamphlet which purports to reveal a nefarious Jewish conspiracy to dominate modern politics. In the early twentieth century it was recognized that the *Protocols* draws heavily upon the *Dialogue*, thus revealing the pamphlet as a forgery. That notwithstanding, the *Protocols* became a favourite of Hitler and is widely cited by extremists who want to reinforce their claims of a 'Jewish conspiracy'. Ginzburg unpicks the many influences upon these two texts, the relationship between them and the extent to which they may reveal something of the reality of the modern political world. In the case of the *Dialogue*, Ginzburg argues that Joly's Machiavelli blurs the boundaries between fiction and history and is intended to represent a real enemy, the modern despotic state under which Joly lived. In the case of the *Protocols*, its historical interest does not end with exposing it as a pernicious forgery: it too can be made to yield information about the political and intellectual circumstances in which it was produced. Ginzburg's study demonstrates that assiduous attention to context can reveal much in ambiguous and seemingly intractable evidence. The most valuable historical sources are not always those which present themselves as such, and accepting sources at face value can have the most appalling results, not just for scholarship, but also for society.

The geologist Vincent Courtillot considers a question which has been central to the development of modern palaeontology, biology and geology: has Earth's history been punctuated by massive catastrophes? In particular, Courtillot asks whether it really was a meteorite strike that brought about the extinction of the dinosaurs. The author brings together a number of different scientific disciplines, each with differing methods and expectations. In dealing with vastly significant yet imperfectly understood events over a timescale of millions of years, Courtillot demonstrates that geology and its related disciplines are essentially historical sciences and, as such, are unlike physics and chemistry, whose techniques they use. In important ways the evidential problems faced by the geologist are similar to those faced by the historian: such material as can be found is often patchy and only indirectly relevant to the problems at hand. As with history, the present-day context may serve as a poor guide to interpreting material dating from a time when the world was substantially different. Courtillot draws on a wide range of different types of evidence – observation of minerals, laboratory experiments, numerical models and inferences from one space and time to another – and emphasizes the importance of keeping track of the assumptions implied by each approach. Courtillot's test of his theory is whether it allows further inferences to be made and explains new discoveries. His theory includes much that was, up until recently, contrary to scientific orthodoxy, but he shows how numerous and disparate pieces of evidence can be woven into a single elegant explanation of Earth's distant past.

Monica M. Grady addresses the subject of life beyond Earth. Her question is not whether or not such life does exist, but rather whether or not it might exist and what would count as conclusive proof that it does. Grady argues that our best starting point for considering these questions is to use our only known instance of life: on Earth itself. She considers the biological envelope of life on Earth – the limits within which life can survive – to indicate what might be possible elsewhere. Life has been identified in remarkably inhospitable places, at temperatures well above the boiling point and well below the freezing point of water, at acidities ranging across almost the whole pH scale and at pressures of up to 100 atmospheres. Grady points out that the surface of Mars shows many similarities to Antarctica, in which a significant biomass of simple organisms, including lichens, have been identified. Similarly, the discovery of successful ecosystems around hydrothermal vents on the ocean's floor, based on chemical energy rather than oxygen and photosynthesis, encourages us to search for life in other deep oceans such as that on Jupiter's satellite, Europa. A basic assumption in Grady's work is that the fundamental chemical and physical rules which govern life on Earth necessarily hold elsewhere in the Universe. This assumption gives access to a large body of empirical evidence (life on Earth) on which to base our theories about less accessible planets. The logical possibility that life may find other loopholes in extreme environments opens up infinite possibilities for theorizing; but prioritizing empirical research exposes definite tracks to follow in the search for intergalactic Woozles.

Brian Greene concerns himself with the most universal of questions. How did the Universe begin? What are space and time? The main problem with establishing a general physical theory of the Universe is that it needs to combine two fundamental theories about the physical world: the Theory of Relativity, which deals with the gravitational forces that work over the immense distances of the Universe; and Quantum Mechanics, which deals with the powerful forces that keep subatomic particles – the building blocks of the Universe – together. Both theories are essential to our understanding of the physical world and both provide phenomenally accurate predictions. These theories, however, are incompatible. Whilst General Relativity provides a geometrical, smooth description of forces, Quantum Mechanics requires particular mathematical discontinuities. Greene believes the resolution to this inconsistency can be found in 'String Theory', a theory that goes beyond our current knowledge about the smallest parts of the Universe. In this theory, subatomic particles are said to consist of even smaller parts, string-like filaments of energy, which can iron out the discontinuity between the world of relativity and the quantum world. The physical consequences of this theory, however, are strange and present us with a Universe of hidden extra dimensions curled up beyond our empirical reach. The strength of this theory lies in its elegance, mathematical consistency and explanatory power. Yet, as Greene admits, there is no experimental basis for the theory, nor indeed does current technology even allow that possibility. Greene expresses the hope that empirical evidence will one day be available; but when grappling with these fundamental problems of physics, evidential priorities are very different from those to be found in the work of, for instance, Grady or Courtillot.

Philip Dawid examines the relationship between statistics and the law, two subjects which at first glance employ different asssumptions and methods and yet both of which attempt to establish means for dealing with incomplete evidence. Statisticians seek to represent the degree of incompleteness of evidence, whereas lawyers are forced to act in the face of it. Dawid's interest in the relationship between these two disciplines is not simply theoretical. In recent years, in a number of high-profile cases, statistical arguments have been advanced in courts of law, most controversially so in the case of trials relating to multiple infant deaths. Such statistical evidence is often treated as a matter of intuition, but it is far from straightforward in its application. Through no malicious intent, statistics have often been misused recklessly in courts. Recent years have seen a number of high-profile exposés which may have discouraged prosecutors and defenders from employing statistical evidence. Dawid argues strongly that we should not be timid in our use of statistics, simply rigorous. This is particularly true given how often certain categories of cases rest on DNA profiling which can never provide a 100 per cent identification, but merely a probability of correct identification. In addition, Dawid points to a number of ways in which statistical and logical principles can be used to weigh the relative importance of different categories of evidence, both in a court and across all fields of human enquiry. It is striking that our explorations of formal logic reach back into antiquity and yet our principles for interpreting and weighing evidence seem relatively underdeveloped.

Where Dawid discusses how courts ought to treat evidence, Cherie Booth inverts the problem. Her essay examines the function of civil and criminal law courts within society and considers how this function influences the law's approach to the problems of evidence. English courts (like many others) are governed by two powerful imperatives: first, the presumption of innocence, which it is the obligation of the prosecution to disprove; secondly, the necessity to reach a verdict even in the face of imperfect evidence. These two imperatives can pull in opposite directions, encouraging both great rigour in, and a certain relaxation of, standards of proof. In the case of the presumption of innocence, the possibility that a defendant is not guilty can be much smaller than might normally be considered statistically significant and yet it can still weigh heavily in the courtroom. In the case of imperfect evidence, courts are often unable to apply rules of probability that apply to the rolling of dice and instead make judgments based on experience, precedent and other less readily definable considerations. In principle, this problematic inconsistency is reconciled by the fact that courts are concerned only with the specific people and particular circumstances before them; they have no concern with establishing general rules for the wider prediction of guilt or innocence. Booth considers how successfully this reconciliation can be effected in practice through an examination of legal proceedings relating to three very different areas which have seen much development in recent years: domestic violence; corporate manslaughter proceedings; and war crimes tribunals.

Karen Armstrong considers the relationship between evidence and religious belief. She criticizes common ways in which questions of evidence are applied to religious belief. For the believer, lack of evidence might stand as a badge of pride, indicating faith; for the unbeliever it is a fundamental weakness. Armstrong argues that in both cases this approach is flawed, arising out of a misunderstanding of the nature of many religious texts and an ignorance of much early teaching. She points to Plato's famous distinction between two different types of understanding: *mythos* and *logos*. *Logos* refers to rational understanding which is the result of intellectual or practical enquiry. *Mythos*, however, is concerned with understanding the truth of those many areas of life which are not altogether subject to rational explanation, such as pain, meaning and destiny. Armstrong argues that subjecting the mythic elements of religion to logical processes of proof results in both bad religion and bad science. In her analysis this is very much a feature of religious thought of the post-Enlightenment world. She sees the same impetus, for example, behind attempts by the Christian Right to discover the remains of Noah's Ark today. By implication, Armstrong equates the rise of scientific understanding and its associated logical processes of reasoning with the rise of religious fundamentalism and its desire to defend the *mythos* of religion as though it were *logos*. For this reason Armstrong simply rejects the relevance of evidence for questions of religious belief.

These eight chapters offer a two-fold impression. They demonstrate powerfully how far we have come in generating knowledge and understanding problems that often fall well beyond our immediate experience. They also show how approaching the blank spaces in our network of knowledge requires careful consideration of our use of evidence. At a time when we are asked to take much on trust and the general public displays a healthy scepticism toward officially sanctioned knowledge, it is especially important to scrutinize the basis of our understanding of the world around us. What is often presented as if it were as incontrovertible as the implication of the footprint on Robinson Crusoe's island can in fact be tracks left by those who want us simply to follow.

1 Evidence and Explanation

PETER LIPTON

Two questions about evidence

The contributions in this collection explore the nature and function of evidence in diverse disciplinary contexts. In this chapter, I approach the topic in general terms, from the characteristically abstract perspective of the philosopher, though I will be promoting a concrete proposal about how we go about determining what the evidence shows. Philosophers worry about how knowledge is possible. Since acquiring knowledge depends on the proper use of evidence, philosophers also lose sleep struggling to understand what evidence is and how it works. And you should be warned that this confused insomnia can be contagious. As the moral philosopher Phillipa Foot once said, 'You ask a philosopher a question and after he or she has talked for a bit, you don't understand your question any more.' That is the sort of risk you run by reading on, since this essay considers some of the answers philosophers give to questions about evidence.

Think of your beliefs taken collectively as forming a vast network. Now think about the evolution of that network over time. It evolves in two different ways: externally and internally. In external evolution, the network grows by the addition of new beliefs from outside. This is what happens in perception, when you see something new. It also happens in testimony, when a belief gets added to the network because someone told you so. Of course you are not entirely gullible. You do not believe everything you hear, and sometimes you do not even believe your own eyes. But perception and testimony are fairly direct sources of new belief, and their source is external to the network. The second type of evolution of belief is internal: here the source of the change in the network lies within the network. For example, you may notice that some beliefs contradict others, so something has got to give. The network contracts, with beliefs dropped in order to restore consistency. But internal evolution can also involve growth, the addition of beliefs, because the old beliefs already in the network themselves generate new beliefs, by inference.

It is this organic growth from within that I focus on in this essay, because this is where evidence figures. For our purposes, we may take evidence to consist of beliefs in our network that generate and direct this growth. The process of generation, of web-spinning, is inference. So to call a belief 'evidence' is to describe it in terms of what it does. Evidence leads to inference, where old beliefs generate new ones. Coming home, I see the dirty dishes in the kitchen sink. That is a case of the external evolution of my network: a new belief from without. But if I go on to use that new belief as the basis for the inference that my kids are home, we have a case of internal evolution, with the dirty dishes as my evidence. Of course we are doing this sort of thing all the time, not just in the home but also in science, in history, in the law, everywhere. Seeing the fossil, the palaeontologist infers a long-dead dinosaur; observing that the galaxy's characteristic spectrum is shifted towards the red, the astronomer infers that the galaxy is receding from us at a particular speed; reading the documents, the historian infers the causes of a war; finding various clues at the scene of the crime, the detective infers the identity of the perpetrator.

Philosophers have been obsessed with two questions about evidence and inference. The first has to do with the direction of the evolution. How do the beliefs already in the network determine which new beliefs will be added to it? How do we decide which way the evidence points? This is a descriptive question, a question about how people actually use evidence. The second question is about justification. Do we use evidence properly? It is no good adding just any belief to the network. There are various virtues a new belief should have, the most obvious of which is that it should be true. The question of justification is the question of whether there is any reason to believe that the way we use evidence makes it likely that we are meeting this basic condition. It is the question of whether there is any reason to believe that the way our network evolves when we make inferences is a way that tends to add truths and not falsehoods.

I will consider both the question of description and the question of justification. Intuitively, one would have thought the description question of how we actually use evidence should be addressed first, before we tackle the justification question of whether that is a good way to use evidence, from the point of view of discovering the truth. For until we have answered the descriptive question, it looks as though we cannot even properly ask the justification question, for we cannot say just what the practice is that we are supposed to justify. Nevertheless, philosophers have tended to focus first on the justification question, and I follow that precedent here, though I will also consider why philosophers have proceeded in an order that apparently gets things back to front.

Hume's problem

Some of the inferences made from evidence already in the network are particularly strong, because the truth of the evidence positively guarantees the correctness of the inference, the truth of the new beliefs generated. This kind of inference is deduction, and it is very nice work when you can get it. For example, if I already believe that there are 22 people in the room, then I may infer with complete confidence that at least 4 of them will have their next birthday on the same day of the week (though not necessarily on the same day). How can I be so confident of this inference? The point is not that it is certain that there really are 22 people in the room; after all, I could have miscounted. The point is rather that if there are 22 people in the room, then at least 4 must have birthdays on the same day of the week. Why is this? Well, there are only 7 days in a week, and in distributing 22 items among 7 boxes, putting at least 4 items in one of the boxes is unavoidable. That is deduction: the truth of the premises, the evidence, guarantees the truth of the conclusion, the inference. If the premises are true, the conclusion *must* be true; equivalently, it is *impossible* for the premises to be true yet the conclusion false.

It is sometimes said that triviality is the price that deduction pays for security, because the only way that it can be impossible for the premises of an argument to be true yet the conclusion false is if the conclusion is already contained within the premises, which is to say that the conclusion says nothing new. But certainly the premises may be present in a person's network of belief without the conclusion already being there as well. Someone could believe that there are 22 people in the room, without believing anything about how many of them will have their next birthday on the same day of the week. So the generation of the conclusion of that argument marks a genuine addition to the network. Indeed conclusions of deductive arguments can be quite unobvious, as we see for example in a good verbal puzzle. A father and his son are driving down a dark road when they have a terrible accident, hitting a big tree. The father is killed instantly; the son is critically wounded. The son is rushed to the hospital and straight into the operating theatre. The surgeon walks into the theatre, takes one look at the boy and says, 'I can't operate on this boy, he is my son.' How is this possible? (No, the 'father' is not a priest, and the puzzle has nothing to do with adoption or exotic genetic technologies.) Many thoughtful people find this puzzle inordinately difficult, yet the answer follows directly by deduction from the information given. The surgeon, of course, must be the boy's mother.

One can raise questions of justification about deduction. For example, how can one know that an argument that seems deductively valid really is so? Sure, it seems that if a person has a son and that person is not the boy's father, then the person must be his mother, but do we *really* know that there is no third option? This sort of sceptical worry – the worry that the evidence might be fine but the inference from it unsound – is, however, strained in the case of deduction, because when we judge an argument to be deductively valid it is not just that we think it likely that the conclusion is true if the premises are true. It is that we cannot even conceive of how the premises could be true yet the conclusion false.

But people do not expand their networks of belief internally by deduction alone: we are and must be more ambitious in the inferences we draw. For deduction will not, for example, enable us to make inferences about what will happen in the future from evidence about what has happened in the past. Nor will it enable us to infer who committed the crime from evidence at the crime-scene. Nor will it enable us to infer dinosaurs from fossils. Nor indeed will it enable us to infer what our beloved is feeling or thinking, from the evidence of her or his behaviour. For in none of these cases does the evidence entail what we infer; in all of these cases, that is, it remains possible that the evidence is as we believe it to be, yet our inference is incorrect. These non-demonstrative inferences are inductive inferences, and we rely constantly on induction, on inconclusive reasons, in science and in everyday life alike.

In the case of induction, the question of justification is acute. The problem is not to show that the beliefs added to the network in this way are invariably correct: we know that this is not so, since we sometimes make predictions that turn out to be mistaken, even though the observations upon which they were based were correct. The problem is, amazingly enough, just to show that our inductive inferences are better, more reliable, than flipping a coin. In other words, the challenge is to show that inductive evidence is not entirely worthless. A pretty modest challenge, one would have thought, and indeed the problem would not be very interesting if it were not for the fact that nobody has been able to solve it.

This problem of induction was made corrosive by the great eighteenth-century Scottish philosopher David Hume, who maintained that no inference that falls short of proof is any better than a blind guess. In other words, Hume argued that when we expand our network of beliefs by inductive inference, as we are constantly doing, we have no reason whatever to think that the new beliefs are going to be true. It is not news that some dead philosopher held an improbable view; but Hume gave a fabulous argument for his view, and it is a point of sublime aggravation that nobody has come up with an adequate reply.

Hume begins by assuming an answer to the question of description, the question of how we actually make our inductive inferences. According to him, when we expand our network of belief by inductive inference, the basic method is simple extrapolation, the basic principle 'More of the Same'. We see a pattern in the world, and we predict that pattern will continue. Unsupported books have always fallen to the ground in my experience, so I predict that they will do likewise in the future. There is no logical contradiction in supposing that they fall in the past but not in the future, yet I would wager almost any amount that this is not the way things will go. To show that this way of using evidence is better than guesswork, however, requires giving some reason to believe that More of the Same is better than guesswork, some reason to believe that it will be at least moderately reliable, tending on the whole to take me to new beliefs that are true. How is that to be shown?

What is needed is some *evidence* for the claim that More of the Same is going to work, going to take me to true beliefs, at least more reliably than blind guessing. Anything I already know without using induction is admissible evidence. Perhaps the most promising evidence is the actual track record of the inductive inferences previously made and checked. This is not the time to feign modesty. My track record has been impressive: much better than guesswork. Otherwise, I wouldn't be here to worry about the problem of induction. In short, my reason for thinking that induction will work in future is that I have observed it to work in the past.

Unfortunately, Hume has a brilliant reply to the appeal to track record: 'So what?' More of the Same certainly has worked so far, but why is that supposed to give me evidence that it will work in the future? The inference would itself have to be deductive or inductive. It cannot be deductive, because the fact that More of the Same has worked in the past obviously does not prove it will work in the future. And to say it has worked in the past so it is likely to work in the future – to give an inductive argument for induction – well, that is like deciding to trust someone because they tell you they are honest. You can't use an inductive argument to justify induction. That really does seem worthless. After all, by parity of reasoning, I suppose someone might defend the method of guessing by saying that it is his guess that guessing will work.

Heads or tails; if it is heads, we lose; if it is tails, we lose. From this it follows, by deduction, that we must lose. Similarly, any reason to trust our way of using evidence in induction would have to be deductive or inductive; there can be no deductive reason to trust induction; there can be no inductive reason to trust induction. From this it follows, by deduction, that there is simply no reason to trust induction. It is not just that nobody has yet found a way to justify induction. If Hume's great sceptical argument is sound, no justification is possible.

From justification to description

Perhaps, as the American philosopher Willard van Orman Quine once wrote, the Humean predicament is the human predicament: Hume's sceptical argument is sound, so we can have no reason to believe one non-demonstrative inference over another. This is literally incredible – it is literally incredible that my computer keyboard is really just as likely to burst into flames right now as not – but the truth is alas not constrained by what we can bring ourselves to believe. There is, however, some room for negotiation over what would count as a solution to Hume's problem. A full and straight solution would be an obviously cogent, circle-free argument showing that our inductive practices are likely to be at least moderately reliable, tending to take us from truths to truths. Faced, however, with the apparent unavailability of such a thing, philosophers have naturally wondered whether there is something weaker – that is, something one could provide – which would go some way towards drawing the sting from Hume's argument.

One approach is to argue that our inductive practices do not have to be shown to be reliable in order for it to be rational to use them or in order for them to be means for the generation of new knowledge. Thus it has been suggested that someone who forsakes even the broad contours of our inductive behaviour – predicting change when everyone else predicts More of the Same, trying to leave by the window when everyone else takes the door – is irrational, whatever happens. Rationality is different from reliability, since a madman might yet turn out to be correct, but it is part of the very meaning of our concept of rationality to use our inductive methods. Thus to ask whether induction is really rational is like asking whether bachelors are really unmarried. To ask the question is to betray a failure to understand the relevant concept, whether of bachelorhood or of rationality. The reliability of our practices cannot be defined into existence, but their rationality is.

Unfortunately, like most of the indirect solutions to the problem of induction, this appeal to the semantics of rationality suffers from the familiar problem of the drunk and the lamppost. A policeman saw a drunk late at night, swaying about next to a lamppost, staring at the ground. When asked what he was doing, the drunk's slurred reply was that he was looking for keys he has lost. 'Did you lose them around here?', asked the policeman. 'No', replied the drunk, 'but the light is so good'. Perhaps making inferences in broadly the way we do is constitutive of the notion of rational thought, but once a sharp distinction is drawn between being rational and being reliable, the natural moral to draw is that Hume's problem really is about reliability, not rationality, and the semantic solution helps here not at all. I do not want to be rational: I want to be right.

A second attempt to circumvent the challenge of justifying our inductive practices is to focus more directly on reliability, rather than attempting to ignore it. Thus it has been argued that what counts for knowledge is not that we can show that our methods are reliable, but just that they are reliable in fact. The application of this externalist, reliabilist view of knowledge is easiest to see in the case of perceptual knowledge. I look up at my office desk and I see the fountain pen I thought I had left at home. I thus come to know that my pen is on that desk, and not because I have established that this could not be a dream, or that I am not really just a brain in a vat floating out in space, or any of the other sceptical fantasies that have plagued the history of philosophy. Rather, I know that the pen is on the desk because I am, not to put too fine a point on it, a reliable pen detector. I truly believe that the pen is there, and if the pen hadn't been there I wouldn't have thought it was, since I wouldn't have seen it. And this picture of knowledge as depending on reliable detection can be extended to inductive inferences. Such inferences yield knowledge when they reliably yield true belief. There is no further requirement. In particular, there is no requirement that the reliability be demonstrated or justified.

Of course if our inductive practices are actually unreliable, we are out of luck. But this is true on any view of the matter. The point of the reliabilist's solution to the problem of induction is just that if, in fact, our practices are moderately reliable, as we all believe, nothing further is required in order for those inferences to yield knowledge. And it is important to note that Hume never tried to argue that induction will be unreliable. He was too smart to try that: he knew that a prediction of failure is as much a prediction as a prediction of success. Rather what Hume argued is that we have no reason to trust induction and it can yield no knowledge, even if it is reliable, since that reliability could never be shown. This is what the externalist denies: he or she insists that the reliability itself is enough. Since it is certainly possible that our inductive practices are reliable, it is possible that we have knowledge by induction too.

This is a better response to Hume than the semantic solution, but it still fails to scratch everywhere it itches. For even if knowledge by induction does not require that we have good reason to believe that induction is reliable – just that it be so – still, we would like to have such a reason, and indeed deep down we believe that we do have one, in the track record of our inductive methods. Surely the fact that induction has worked in the past does give some reason to trust it in future. And yet Hume's observation that this begs the question – that you cannot use induction to justify induction – seems devastating. Yet the reliabilists can take their approach one step further, to defend exactly such an appeal to track record. For suppose that someone began with blind faith in the reliability of induction. Even if induction is in fact generally reliable, this faith would not count as knowledge, since it would not itself have been acquired in a reliable way. Suppose now, however, that the reliabilist surveys the track record of her methods, and finding them to have been reliable in the past, uses this as a reason for the claim that induction will also be reliable in future. An inductive argument, to be sure, but one that would, if induction is in fact reliable, take her from blind faith to knowledge, on the reliabilist's conception. Such an argument would still be circular in the sense that it would never convince a die-hard inductive sceptic, but it would have cognitive value for those willing to use it, since it would take such people from mere belief to knowledge of the future reliability of induction.

That is as far as the reliabilist will take us. A residual itch remains, since we would really like an argument for our methods that blows the sceptic out of the water. Still, in my view the reliabilist takes us as far as any solution to Hume's problem that has been proposed. And perhaps that is just because the residual predicament – the inability to refute the sceptics on their own terms – is simply the human predicament.

Hume's own response to his problem is different. He gives what he aptly calls a 'sceptical solution', a solution which does not blunt the force of his sceptical argument but quite consciously shifts focus from the problem of justification to the problem of description. Having argued that our inductive inferences are unjustifiable, and hence could not be governed by reason, the natural question becomes the question of what then does govern them. Hume's answer is Pavlovian (a century before Ivan Pavlov was born): we are habit-forming creatures, and having observed one thing to follow another, we come to expect things of the second sort when we see things of the first sort. The principle of More of the Same has no justification, but it operates nevertheless, as what Hume calls 'a natural instinct'.

However much one might hope for something closer to a straight solution to the sceptical argument, the descriptive question which Hume turns to answer is certainly legitimate. And, as I noted above, one would have thought it prior to the question of justification, since it specifies what it is that we are supposed to justify. This intuitive order is reversed in Hume for two reasons. First of all, he does not need a careful description of our inductive practices in order to prosecute the sceptical argument, because that argument only depends on the fact that inductive reasoning, whatever form it takes, falls short of deductive validity. Secondly, as we have seen, he uses the failure to justify induction as a motivation to give a psychologistic answer to the descriptive question.

In any event, the descriptive question is of considerable interest whenever it arises. How do we in fact use evidence, for better or for worse? Hume's particular answer, however, is not very good as it stands. More of the Same cannot be our only inductive principle, since we do sometimes predict that things are going to change. And, especially in science, there are many inductive inferences where what is inferred – such as invisible particles and processes – is unobservable and not just unobserved. Here Hume's Pavlovian habit mechanism gets no purchase, since the expectation can never be reinforced – it is as if the dog never knew whether or not it was fed after the bell rang. For reasons I have just suggested, the weakness of Hume's description does not alas point to a flaw in his sceptical argument; but it does offer us the challenge of providing a somewhat better description of our 'natural instincts'.

The challenge is surprisingly difficult to meet. That the justification of our practices should be a tricky business is perhaps to be expected, but if you want simply to know how a scientist goes about weighing evidence, then why not just ask him or her? This reaction, however, seriously underestimates the gap between what we can do and what we can describe. It is one thing to know how to tie one's shoes or to ride a bike; it is quite something else to be good at describing the motions, physics and physiology that these activities involve. The contrast is even more striking in the case of cognitive abilities. To be good at discriminating grammatical from ungrammatical strings of words in one's native tongue is a far cry from being able to describe the principles or mechanisms one employs to that end. It is the same with evidence. Scientists may be very good at weighing evidence and at making sage inferences, but they are very bad at saying how they do it. This is no criticism of science: epistemology is not their business. What is more embarrassing is how difficult philosophers of science have found the task – more embarrassing, because this is their business. Embarrassing or not, that is the task for the remainder of this chapter. What I will suggest is that we can make some progress on the descriptive problem by looking first not at evidence at all, but at explanation.

What is explanation?

It is one thing to know that something is the case, another to understand why. We all know that the sky is sometimes blue, but few people understand why. Surely it ought to be almost entirely black! We all know that the same side of the Moon always faces Earth, but most people if asked will say this is because the Moon does not spin on its own axis. This is a mistake: if the Moon did not spin, then different sides would face Earth as the Moon orbits Earth. (At this point, readers may wish to deploy a couple of handy objects for purposes of simulation.) In order for the same side of the Moon always to face Earth, the Moon must spin around its own axis, and moreover that period of the spin of the Moon around itself must be exactly the same as the period of the Moon's orbit around Earth. (The axis of spin must also be just perpendicular to the plane of orbit.) This is a remarkable coincidence, crying out for explanation, an explanation that most people cannot give. That is, most people know that the same side of the Moon always faces Earth but do not understand why.

Explanations are what give us that understanding. And we are obsessed by explanation, constantly asking and trying to answer why-questions. Once the distinction between knowing that and understanding why is foregrounded, this can seem a surprising obsession. Why don't we just stick to the facts? Why isn't it enough to know that something is a certain way, without having to understand why it is that way? (Notice that these two previous sentences are themselves why-questions.) Indeed sometimes we mystify ourselves through this obsession, looking for explanations of the wrong sort, based on diverse misapprehensions. So we have the story of the paranoid man who sees a map of London in King's Cross with a big red arrow on it, saying 'You Are Here'. 'How did they know?', he worried. Sometimes the correct response to a request for an explanation is not to give a direct answer, but to show that the question was misguided. But most why-questions are perfectly legitimate. Indeed, you can sensibly ask 'why' about almost anything, including the answer to another why-question itself, as you probably discovered many years ago. Why does the same side of the Moon always face the Earth? Because its period of spin is the same as the period of its orbit. Why is that? Because the Moon is slightly oblong. Why does being oblong lead to linked periods? And so on. This regress of whys is a fundamental feature of explanation that most of us discovered as young children, to our parents' annoyance. The moral of the regress is not that explanation is futile, but rather that what explanations provide – understanding – is not like a substance that is transferred from explanation to phenomenon explained. For what the regress shows is that B may explain A, even though B has not itself been explained.

The central task of a philosophical account of explanation is to say what the relationship between two things must be, for one to explain the other. What is the explanatory relation? This is a large question, but in many cases the answer is that we explain phenomena by citing their causes: the explanatory relation is the relation between cause and effect. One of the attractions of a causal theory of explanation is that it accommodates the 'why' regress we have just described, since we may know that B caused A without knowing what caused B. At the same time, the causal theory must be incomplete, since there are some good explanations that are not causal. For example there can be explanations in pure mathematics, an area where there are no causes to be found. Thus one can explain why the area of one square will be half the area of another square built on the diagonal of the first, but this explanation will not cite causes. And even physical phenomena may have non-causal explanations. Numerous sticks are in free fall, twisting and tumbling. A snapshot is taken before any of them hit the ground, and what is found is that at that moment more of the sticks were near the horizontal than near the vertical. What explains this physical distribution? Geometry provides a particularly lovely explanation. Think of a single stick with a fixed midpoint. There are only two ways it could be vertical – pointing up or pointing down – but many ways it could be horizontal – any orientation in the horizontal plane. More sticks were near the horizontal because of the geometrical fact that there are more ways for sticks to be near the horizontal. But geometrical facts are not causes.

There are two other general aspects of explanation that I will flag here, before returning to the question of evidence. One of these is that the why-questions we ask are very often contrastive. The questions are not just of the form 'Why P?', but 'Why P rather than Q?' (though often the contrast is implicit). Like the why-regress, the fact/foil contrastive structure of many why-questions is a fundamental feature of our explanatory practices that many of us discovered as young children. Why do birds fly south in the winter? Because it is too far to walk. At the risk of spoiling this wonderful joke by explaining it, I observe that it depends on a tacit switching of contrasts. The implicit contrastive structure of the question is 'Why do birds fly south rather than stay where they are?'; but what is answered is a different question: 'Why do birds fly south rather than walk south?' The fact is the same in both cases, but the foil varies, and that makes a difference to what counts as an explanatory answer. In the case of many causal explanations, the foil selects for a cause that marks a difference between the two cases. Thus while getting a first in philosophy requires both ability and effort, only her ability explains why Jane rather than Joe got a first, since Jane and Joe made a similar effort in the course, though Joe has less ability. On the other hand, only her effort explains why Jane rather than Sam got a first, since Jane and Sam have similar ability, though Sam was lazy.

The final feature of explanation I mention here is the prevalence of what are known in the business as 'self-evidencing explanations'. A number of years ago, when I lived in northwestern Massachusetts, a visiting friend asked me to explain the peculiar tracks in the fresh snow she saw one morning in my garden. I replied that someone had gone across my garden on snowshoes. This was a perfectly good explanation, even though as it happens I did not see the person on snowshoes: I just recognized the tracks. Here the person on snowshoes explains the tracks, even though at the same time (and this is what makes the explanation self-evidencing) seeing the tracks was my evidence for the explanation. There is a kind of circularity in self-evidencing explanations – B explains A while A is evidence for B – but this circularity is very common and entirely benign. As we will see, self-evidencing explanations give us a valuable clue about the role of explanation in answering the descriptive problem of inference, in showing how evidence works.

Explanation as a guide to inference

Inference and explanation can both be seen as relations between a pair of statements. Thus A may be evidence for B and B may explain A. But which comes first? The natural thought is that inference comes before explanation. First we make our inferences, then when asked a why-question we look to the pool of inferences we have made to see whether an explanation is to be found therein. If so, we give that explanation; if not, we make more inferences until we are in a position to explain. Thus if you ask me why the same side of the Moon always faces the Earth, I will give you the explanation if I already know the facts that will do this job; otherwise I will have to do further research and inference, and then get back to you with the explanatory goods.

As obvious as this order may seem – inference first, explanation second – things often work the other way round. This is so because explanation is an important guide to inference: the search for explanations directs the way we expand our network of beliefs, and it does so because we very often infer something precisely because it would explain. This may seem at first incoherent, because what is proposed is that the explanation is what we infer to: it is what we add to the network of beliefs. So how can it guide its own creation? Surely final causes have no place, least of all in a Darwin Lecture! But the proposal is not incoherent, and it does not involve anything bringing itself into existence. To see this, we need the distinction between potential and actual explanation. A potential explanation is a hypothesis that will explain, if it is true; an actual explanation is a potential explanation that is true. So all actual explanations are also potential explanations but not conversely, since there will be some potential explanations which would have explained if only they had been true, but alas they are not. 'The dog ate it' is a potential explanation for the phenomenon of a student who turns up for a supervision without an essay – but it is not always the actual explanation.

The distinction between potential and actual explanation shows how explanatory considerations can take us from evidence to what we infer, even though what we infer is itself an explanation. The picture is this. We start with our evidence, and then before we infer we construct a potential explanation of that evidence, a just-so story, a hypothesis. Then we ask how well that hypothesis would explain our evidence. If the potential explanation is good enough, we infer it. That is, we infer that a good potential explanation is an actual explanation, and so add it to our web of beliefs. That is the idea behind the philosophical account of evidence known as explanationism, or 'Inference to the Best Explanation'. Notice how closely related this is to the phenomenon of self-evidencing explanations. The tracks in the snow are evidence for the person on snowshoes that explains those tracks. Explanationism takes this idea, generalizes it, and adds the following thought: the tracks in the snow are evidence for a person on snowshoes *because* the existence of a person on snowshoes would explain those tracks. That is the sense in which explanation comes before inference and the sense in which explanation is a guide to inference.

The explanationist idea is that we often work out what the evidence shows by asking what would explain that evidence. That is, when we make an inference from our current network of beliefs, we often decide which direction the expansion will go by considering which potential expansions would explain the evidence in the network. And when we consider more than one potential explanation, we often select the potential expansion that would give the best explanation. But what makes one explanation better than another? This is a large question, and one that has only been very partially answered by those of us who worry about how explanations work, but there is one relevant distinction that we can draw here. That is the distinction between the potential explanation that is likeliest to be true, and the potential explanation that would if correct provide the most understanding – the 'loveliest' explanation. Faced with competing explanations vying for a place in our network of beliefs, we want to select the likeliest, because we want to infer truths. The question is how we judge one explanation to be likelier than another. According to an ambitious version of Inference to the Best Explanation, we do this by means of judgments of relative loveliness: loveliness is our guide to likeliness.

In other words, what we do is to infer that the loveliest potential explanation is likely to be an actual explanation, and so we add that explanation to the network.

Explanationism provides a partial answer to the descriptive question about inference, an answer according to which evidence is what is explained, and the direction we take from the evidence is determined in part by which direction would leave us with the best explanation of our evidence. On this account, evidence and explanation are very often partners, part of the same relationship: we infer from evidence to explanation, and we make the inference because the explanation would explain the evidence. That sort of description fits the cases I mentioned at the beginning of this essay. My kids being at home would explain the dirty dishes, the dinosaur would explain the fossil, the movement of the galaxy would explain the red-shift, the historical events would explain the content of the documents, the guilt of the butler would explain his fingerprints on the gun. And we make these inferences precisely because those hypotheses would explain the evidence.

When I introduced the idea that there is a sense in which an explanation might guide its own inference, I worried about an illicit appeal to final causes, something Darwin would have particularly disdained. I hope it is now clear why nothing of the sort is involved, so Darwin can rest easy. And indeed, far from turning in his grave, Darwin in fact embraced explanationism. For in the final chapter in the sixth edition of *On the Origin of Species* he wrote that:

> it can hardly be supposed that a false theory would explain, in so satisfactory a manner as does the theory of natural selection, the several large classes of facts above specified. It has recently been objected that this is an unsafe method or arguing; but it is a method used in judging of the common events of life, and has often been used by the greatest natural philosophers.

Psychological plausibility

How good is Inference to the Best Explanation as an answer to the descriptive question about evidence? I have here only been able to sketch the outlines of this approach. There is a lot more to say, for example about what makes one explanation better (lovelier) than another, though only some of this has so far been said by those of us working in this area. And there are sure to be aspects of our inferential practices that explanationism will not cover. (The impact of negative evidence on the network is one important area where explanationism will require supplementation.) At the same time, you will have gathered that explanationism strikes me as a promising approach to understanding how evidence works. It accounts for the fact that we often ask why the evidence is as we find it to be in order to work out what we should infer from that evidence. It captures our practice of generating just-so stories or simulations, of constructing possible explanations in order to decide which of them are actual explanations. But the strongest argument for explanationism is probably just that it gives a natural description of so many of the inferences we actually make.

Inference to the Best Explanation also comes out well when compared with other accounts of evidence, including accounts that purport to say how we should infer, not just how we actually do infer. These accounts – such as Mill's Methods, Hypothetico-Deductivism and Bayesianism – are broadly compatible with explanationism. Moreover, explanationism can be shown to complement them in various constructive ways, whether by extending their range, by avoiding some of the counterexamples they face, or by supplying a plausible psychological mechanism for realizing an abstract algorithm. Interestingly, we can also find evidence for the psychological reality of explanationist thinking by looking at cases where we tend to use evidence *badly*. Some of these pathologies of reasoning suggest that we are so eager to engage in causal-explanatory inference that we will infer explanations even when we should stay our inferential hand. We crave explanation, and we find a hypothesis attractive just because it seems it would explain, which is just what one would expect if explanationism were correct.

A number of pathologies of inference have been studied empirically by Daniel Kahneman and Amos Tversky. One famous example they discuss concerns Linda the cashier. Having been told that Linda was politically active on the left as a student, the great majority of subjects judged it less likely that she went on to become a cashier than that she went on to become a cashier and active in the feminist movement. This preference violates an elementary fact about probability: a conjunction cannot be more likely than one of its conjuncts. This error in reasoning can, however, be accounted for in part because the conjunctive hypothesis, though it must be less likely than the first conjunct, at least exhibits some explanatory connection between Linda's background and her subsequent behaviour. More generally, a rich and detailed explanatory story may seem more likely, even though the additional detail in fact makes it less likely. And this is what we should expect if we are strongly inclined towards explanationist thinking.

Another striking case that Kahneman and Tversky discuss concerns praise and blame. Suppose that someone is praised for performances that are for that individual above par, and criticized for performances that are below par. What not uncommonly happens is that the performance subsequent to criticism tends to be better, whereas the performance subsequent to praise tends to be worse. (Parents may be familiar with this phenomenon.) Faced with such a pattern, people are inclined to infer that criticism is more effective than praise. This is a natural enough thought, but it is unwarranted by the evidence, since the observed pattern is just what one would expect if neither praise nor criticism had any effect at all. For even without any praise or blame, below-par performances will tend to be followed by something a bit better, above-par performances will tend to be followed by something a bit worse: that statistical fact is what is known as 'regression to the mean'. Yet we so crave a causal explanation of striking patterns that we infer the effectiveness of blame and the counter-productiveness of praise, for this would give a causal explanation of the observed regression, even though in fact such regression requires and supports no such explanation. Cases like this suggest that our inclination to infer explanations of our evidence runs very deep indeed. Whatever one has to say about justification, explanationism seems to capture an important aspect of the way we actually use evidence.

Conclusion

From a certain point of view, it is remarkable that we have so many beliefs. Those beliefs are about a world we inhabit, yet our actual and direct contact with the world is minuscule: a few years, a few places. How do we manage to generate a network of belief that so outstrips even our collective experience? Part of the answer to that question is the process of the organic growth of our network of belief. That process consists in using bits of the network as evidence to guide the expansion of the network. What I have suggested in this essay is that the way the expansion works is often by the construction of trial additions that would explain parts of the network already in place. Thus evidence is evidence for what would explain it, and if an explanatory hypothesis would explain the evidence well enough, it earns a place in an expanded network, perhaps itself to serve in future as the evidential basis for still further expansion.

This is a rather different picture of the way evidence and inference work from that which Hume drew with his principle of More of the Same. Inference to the Best Explanation allows a better description in my view, but it does not avoid the question of justification that Hume made so difficult to answer. We have just seen that explanationist thinking may mislead us in certain cases, but generally it has stood us in good stead, both in science and in ordinary life. That, however, does not settle the question of its future prospects. It seems that we have a reason to claim that inferences to the best explanation are generally reliable, since that reliability would itself explain the conspicuous successes that we have enjoyed with our methods so far. Many of our scientific theories are powerful tools for prediction. If we arrived at these theories through something like Inferences to the Best Explanation, the reliability of this way of thinking would explain why those theories have been so successful. Unfortunately, using an Inference to the Best Explanation to justify Inference to the Best Explanation is no more reassuring than arguing that More of the Same will be reliable in the future because it has been reliable in the past. Explanationism has many virtues, but it does not appear to solve the problem of justifying induction. Probably nothing does.

What looking at explanation does achieve is an illumination of the way we actually handle evidence. If you have made some striking discoveries you cannot explain, and I tell you a story that would make really good sense of them, and a story that would fit well with everything else you already accept, you will find my story almost irresistible. For us, evidence is what we explain, and our explanations are what we infer from that evidence, because they are satisfying explanations. That is the way we think with evidence; indeed it is itself a good explanation.

FURTHER READING

Bird, A., *Philosophy of Science*, London: UCL Press, 1998.

Howson, C., *Hume's Problem: Induction and the Justification of Belief*, Oxford: Oxford University Press, 2000.

Hume, D., *An Enquiry Concerning Human Understanding*, ed. T. Beauchamp, Oxford: Oxford University Press, 1999. [Originally published 1748.]

Kahneman, D., Slovic, P. and Tversky, A. (eds.), *Judgment under Uncertainty: Heuristics and Biases*, Cambridge: Cambridge University Press, 1982.

Lipton, P., *Inference to the Best Explanation*, 2nd edition, London: Routledge, 2004.

Okasha, S., *Philosophy of Science: A Very Short Introduction*, Oxford: Oxford University Press, 2002.

Papineau, D. (ed.), *The Philosophy of Science*, Oxford: Oxford University Press, 1996.

Quine, W. V. and Ullian, J. S., *The Web of Belief*, 2nd edition, New York: Random House, 1978.

2 Representing the Enemy: Historical Evidence and its Ambiguities

CARLO GINZBURG

The word 'evidence' is a piece of historical evidence in itself. In writing about rhetoric, Cicero and Quintillian argued that one of the attributes of a good speech was *evidentia*: the ability to make a topic not only evident, but palpable, 'vivid'. *Evidentia* was a translation of the Greek *enargeia*: a word used to praise historians, poets and painters, since all seemed to share an uncanny capacity to use words or colours to conjure up absent realities, belonging either to the past or to the realm of fiction. In Homer, the related adjective *enarges* is significantly associated with superlative presence: the presence of the gods. Polybius, the Greek historian of the Roman Republic, stressed the truthfulness of the Catalogue of the Ships in the *Iliad*, referring to its *enargeia*.

In ancient Greece, the historian's *autopsia*, or direct visual experience of an event, was conveyed by stylistic 'vividness'. This is a far cry from what we are used to calling 'evidence', which is invariably based on an act of inference. Historians (like judges or policemen) make more or less reasonable inferences about events they did not witness, relying upon evidence as diverse as newspaper articles, fragments of pottery or cigarette butts. But notwithstanding the obvious differences, a certain continuity between the present and antiquity is undeniable, since both our notion of evidence and the Latin *evidentia* emerge in the sphere of rhetoric, especially judicial rhetoric. This may sound odd, since we, after Nietzsche, are used to opposing rhetoric and evidence; but from Aristotle onwards evidence and proofs were a substantial part of rhetoric.

Thirty years ago, Ernst Gombrich published an essay entitled 'The Evidence of Images'. One of the examples he discussed was a word Winston Churchill wrote in the margin of a typewritten draft of the Atlantic Charter. Many

I am very grateful to Michele Battini, Pier Cesare Bori, Cesare G. De Michelis, Andrea Ginzburg, Mikhail Gronas and Sergei Kozlov for their help and advice.

alternative readings could be suggested for this word – a barely readable scribble – but choosing among them would be impossible without access to the context, in the most literal sense, into which the handwritten word had to fit. Then it would be possible to decipher without too much effort Churchill's suggestion of an alternative for the word *because*: *since*.

Gombrich used this example to argue, first, that historical evidence is always more or less ambiguous, and second, that the range of its possible meanings can be narrowed down by checking context. Both conclusions are unobjectionable; today they are more timely than ever. But in order to make his point Gombrich chose a deliberately simple case. If we are confronted not with single words but with longer texts, as is usually the case, a plurality of contexts emerges; since ambiguity affects both the interpreter and the actors, both research and action turn out to involve an intricate network of paths. To make things more difficult (and more interesting) I have selected a case study involving a fictional text and a forgery. This will give us an opportunity to watch the complexity of historical evidence, and the inferences we may make from it.

I will start with a book first published anonymously in Brussels in 1864: *Dialogue aux Enfers entre Machiavel et Montesquieu ou la politiquer de Machiavel au XIXe siècle* (A Dialogue in Hell between Machiavelli and Montesquieu or, Machiavelli's Politics in the 19th Century). The author, Maurice Joly, identified himself on the book's title page only as 'a contemporary'. One year later his true role was ascertained by the French police, whereupon he was put on trial and sentenced to fifteen months in prison for having made seditious and offensive remarks about Napoléon III. The *Dialogue* was immediately translated into German, then republished twice in Brussels in 1868: no longer a mere 'contemporary', Joly was acknowledged as the book's author in these editions. After the collapse of the Second Empire, Joly, a lawyer, launched a political career which soon ended disastrously. In 1878, not yet fifty-five years old, he committed suicide.

A Spanish translation of the *Dialogue in Hell* appeared in Buenos Aires in 1898; then the book fell into oblivion. Two decades later it was identified as the main source of the *Protocols of the Sages of Zion*, the infamous anti-Semitic pamphlet first published in Russia in 1903. The worldwide diffusion of the *Protocols* that followed, still very much alive today, obscured the significance of the *Dialogue in Hell*. Its perceptiveness has been often praised, but only recently

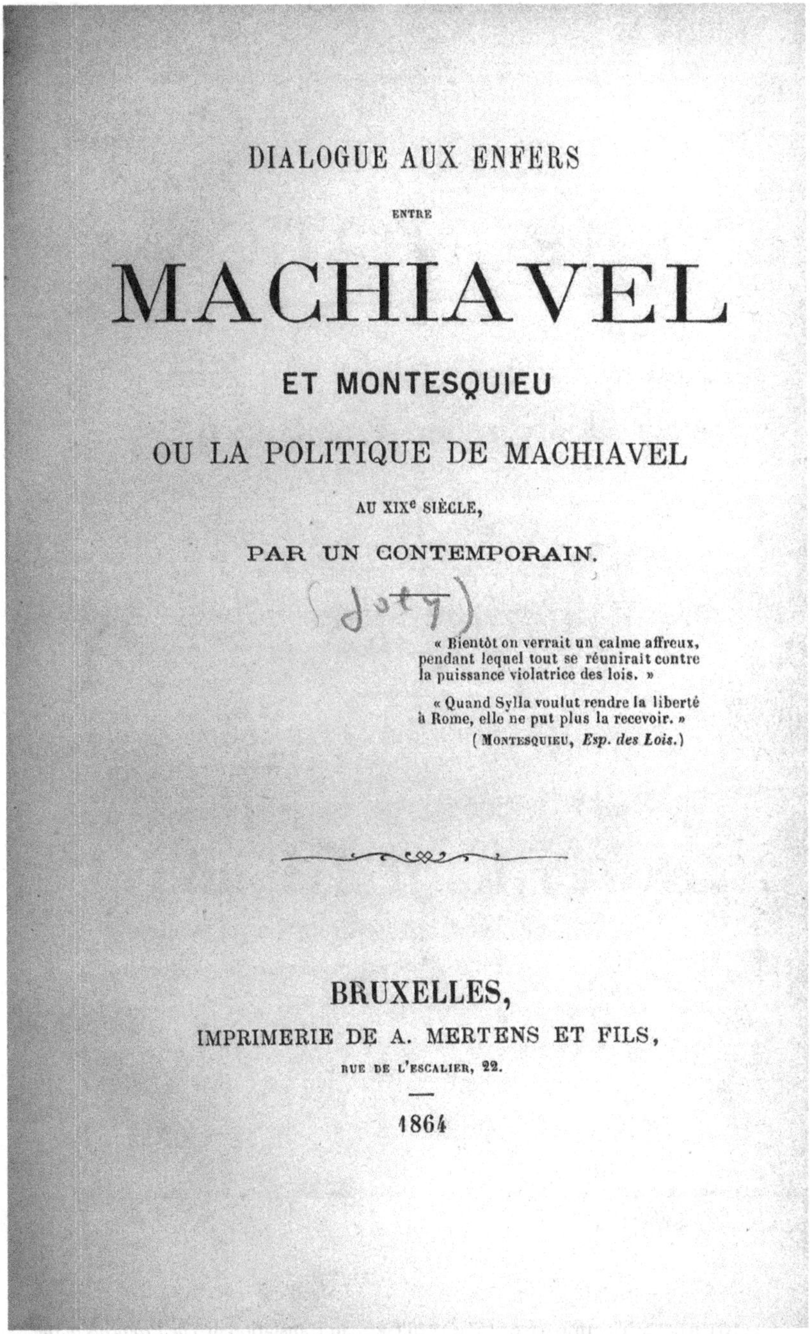

DIALOGUE AUX ENFERS

ENTRE

MACHIAVEL

ET MONTESQUIEU

OU LA POLITIQUE DE MACHIAVEL

AU XIXᵉ SIÈCLE,

PAR UN CONTEMPORAIN.

(Joly)

« Bientôt on verrait un calme affreux,
pendant lequel tout se réunirait contre
la puissance violatrice des lois. »

« Quand Sylla voulut rendre la liberté
à Rome, elle ne put plus la recevoir. »
(MONTESQUIEU, *Esp. des Lois.*)

BRUXELLES,

IMPRIMERIE DE A. MERTENS ET FILS,

RUE DE L'ESCALIER, 22.

1864

FIGURE 2.1. The title page of the first edition of Joly's *Dialogue*.

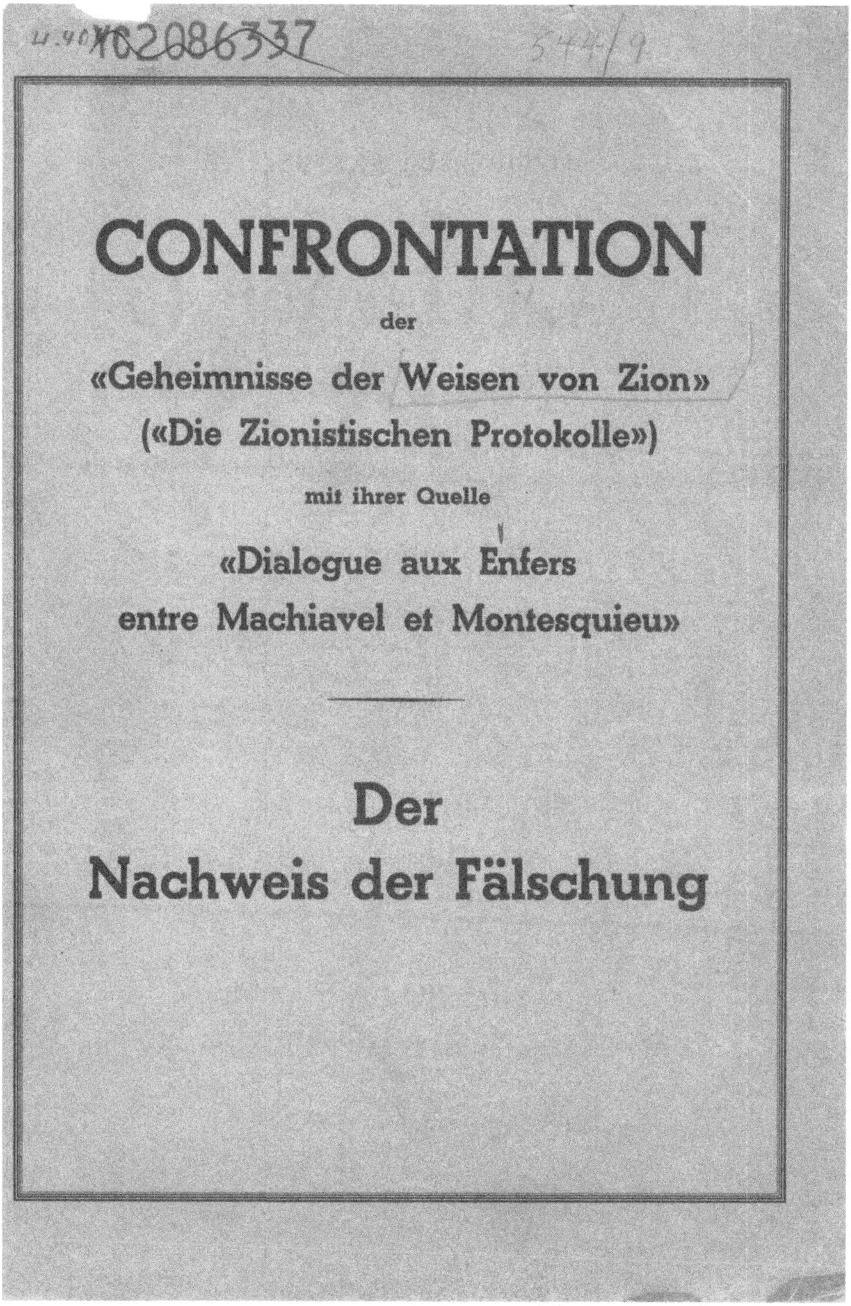

u.40 X02086337   544/9

CONFRONTATION

der

«Geheimnisse der Weisen von Zion»

(«Die Zionistischen Protokolle»)

mit ihrer Quelle

«Dialogue aux Enfers

entre Machiavel et Montesquieu»

Der
Nachweis der Fälschung

FIGURE 2.2. The title page of the 1920 German translation of *The Protocols of the Sages of Zion.*

has Joly's book been rediscovered, especially in France, as a major work – a 'classic' – of nineteenth-century political thought.

I will deal with the reception of Joly's book later. Let us begin with the account of its genesis given by Joly himself, in an autobiographical piece he wrote in 1870:

> One evening, as I was walking on the terrace beside the river near the Pont Royal (much rain and mud, as I still recall), suddenly the name of Montesquieu occurred to me, embodying a whole side of the ideas I wanted to express. But who would be his interlocutor?
> An idea leapt from my brain, and by God it was Machiavelli!
> Machiavelli, the representative of the politics of force, beside Montesquieu, who would represent the politics of law; and Machiavelli would be Napoléon III, who would describe his own abominable politics.

This programme was apparent to the police and to the judges who sentenced Joly to jail. On the basis of the convergence between their reading and Joly's explanation we might conclude that the meaning of the *Dialogue* is unambiguous. But if we move closer still to the text it will tell a different story.

Literary critics have taught us to be sceptical about authorial intentions: they are relevant, of course, but the author is not necessarily the most authoritative interpreter of his or her work. Maurice Joly is a case in point. The first question to be asked about the *Dialogue in Hell* concerns the literary genre (or subgenre) to which it belongs. Joly's recollection shows that the names Machiavelli and Montesquieu occurred to him only once he was already committed to the idea of writing a dialogue. He tells us that he had been inspired to do so by Galiani's *Dialogue sur le commerce des bleds* (Dialogue on the Grain Trade) published anonymously in 1770 and reprinted several times. But the alleged connection between the two books, echoed by all commentators of Joly's *Dialogue*, is unconvincing. In his witty pamphlet, Galiani, a Neapolitan *philosophe*, counterposes his own mouthpiece, a cavalier named Zanobi, with two unknown individuals, one of whom is identified only by his initials. Elsewhere Joly alluded to the *Satire Ménippée*, a late sixteenth-century anti-Catholic pamphlet, partially inspired by Lucian of Samosata, suggesting *it* was a model for the *Dialogue in Hell* – this sounds more plausible. In fact, Joly's fictitious dialogue between two famous political thinkers immediately evokes, both in its title and its content, a literary genre made famous (and possibly invented) by Lucian of Samosata in the first century after Christ – that is, the dialogues of the dead. But now that we have set Joly's *Dialogue in Hell* in its proper niche, its originality appears more striking.

A genre is defined by a series of traits that imply both constraints and possibilities. Obviously, these traits – the so-called laws of a genre – can be violated or modified. If Lucian, inspired by Plutarch, juxtaposes Hannibal and Alexander, he also does something Plutarch never did, setting mythological figures beside his historical figures: Minos sits as judge over the debate between the two great generals in *Dialogues of the Dead*, giving to their comparison a fictional twist. In the late seventeenth century Fontenelle, in his *Nouveaux dialogues des morts* (New Dialogues of the Dead), got rid of mythological figures and focused on historical characters: by effectively reinventing (and reshaping) a genre he was able to stress, in a light, ironic tone, the superiority of the moderns over the ancients. This literary formula rapidly spread all over Europe, from France to England, from Germany to Russia. Joly, who must have been familiar with Fontenelle's *Nouveaux dialogues des morts*, gave the genre another unexpected twist.

The debate in hell between Machiavelli and Montesquieu stretches over twenty-five dialogues (plus an epilogue, written at a later date, which has only recently been appended to the main text). Montesquieu opens by recalling the ideas he expressed in *L'Esprit des lois* (The Spirit of Laws), noting first the autonomy of the three powers (legislative, executive and judiciary). He takes for granted the triumph of this principle, which is going to become the cornerstone of modern states all over Europe – but his information does not go beyond 1847. Machiavelli, with obvious glee, provides Montesquieu with a résumé of French history subsequent to that date: the 1848 revolution and its bloody sequel; Louis Napoléon's *coup d'état* on 2 December 1851; the plebiscite and the proclamation of the Second Empire the following year. To sum up, Machiavelli says, in one of Europe's most civilized countries, disrupted by social and political conflict, one man seized power by force. The new regime effectively combined social peace and prosperity: the most appropriate solution to the fragility of modern society. Here is Machiavelli's eloquent apology for the regime of Napoléon III:

> I don't see any salvation for such societies, veritable colossi with feet of clay, except by instituting extreme centralization, placing all public power at the disposal of those who govern. What is needed is a hierarchical administration similar to that of the Roman Empire, to regulate with machine-like precision the movements of the individual. It calls for a vast system of legislation that

minutely reclaims all the liberties that had previously been imprudently bestowed – in sum, a gigantic despotism capable of striking immediately and at any time all who resist and all who complain. I think the Caesarism of the late Empire answers fairly well to what I desire for the well-being of modern societies. I have been told that such vast apparatuses already exist in more than one country in Europe, and thanks to them, these countries can live in peace, like China, Japan, and India. We must not look down upon these oriental civilizations because of some vulgar prejudice: with each passing day one comes to a better appreciation of their institutions. The Chinese, for example, are very good businessmen and their government is exemplary.

To the readers of the *Dialogue in Hell* Machiavelli's language was self-evident. In 1850 Auguste Romieu had introduced the word 'Césarisme' to describe a regime that was 'the necessary outcome of a stage of extreme civilization . . . neither a monarchy nor an empire, neither despotism nor tyranny, but something specific and not yet very well known'. In the following year Romieu wrote *Le spectre rouge de 1852* (The Red Ghost of 1852), a pamphlet that presented the imminent *coup d'état* by Louis Napoléon as the only way to avert a revolt by the lower classes. Romieu praised force and spoke dismissively of natural rights: 'I believe in social needs, not in natural rights. The word RIGHT is in my view meaningless, since nowhere can I see any equivalent of it in nature. It is a human invention'. Joly's Machiavelli echoes Romieu on this point: 'All sovereign powers find their origin in force, or, what is the same thing, in the negation of right . . . Moreover, don't you see that this word – "right" – is infinitely vague?' But he goes on to associate bluntly 'Césarisme' with a 'gigantic despotism' ('despotisme gigantesque'). This was a direct challenge to his fictitious counterpart, Montesquieu – and to the real Montesquieu as well, who had presented oriental despotism as the very antithesis of progress, which he saw as embodied in European civilization. Joly presumably recalled Tocqueville's rueful reflections on the future of democratic societies: a new form of servitude, 'uniform, mild and calm', mixed with 'some exterior form of liberty . . . protected by popular sovereignty'. But Tocqueville still regarded the freedom of the press as the strongest antidote to the evils of equality. Joly, who had experienced Napoléon III's Second Empire, had developed a different view on this matter. His character, Machiavelli, argued that the best government for modern societies was a form of despotism (we might call it 'occidental despotism') that would pervade all sections of society and coexist with a parliamentary system and a free press: 'One of my great principles', says the fictitious Machiavelli:

'is to set things against themselves. Just as I use the press against the press, I would use oratory to counter oratory . . . In the Assembly, I would control nineteen out of twenty men, all of whom will follow my instructions. In the meantime, I would pull the strings of a sham opposition, clandestinely enlisted to my cause.'

The outcome of this strategy would be, remarks Montesquieu, 'the annihilation of parties and the destruction of other collective forces', although political freedom would formally survive. Machiavelli agrees. He plans to use a similar strategy with the press:

> I foresee the possibility of using the press to neutralize the press. Because journalism wields such great power, do you know what my government would do? It would make itself a journalist; it would become journalism incarnate . . . Like the god Vishnu, my press would have a hundred arms, and these arms would stretch out their hands throughout the country, delicately giving form to all manner of opinion. Everyone will belong to my party without knowing it. Those who think they are speaking their language will be speaking mine. Those who think they are stimulating their party will be stimulating mine. Those who think they are marching under their own banner will be marching under mine.

A perturbed Montesquieu mutters, 'Are these ideas possible or only wild fantasies? They make the head swim.'

Confronted with Machiavelli's lucid arguments and stringent logic, Montesquieu is buffeted between surprise and horror, indignation and bewilderment. He is a man of the past, while Machiavelli is a man of the present and, possibly, of the future. The paradoxical inversion of their relative positions in history suggests a subversion of the meaning often attached to the genre 'dialogues of the dead' since the time of Fontenelle, and more generally a sarcastic rejection of the idea of progress. But Joly's deft use of the dialogic form renders his own attitude somewhat elusive. One is tempted to ascribe a deeper meaning to Joly's claim to have effaced himself as an author – a claim advanced as an apology for the anonymity (ultimately defeated by the Emperor's police) of the *Dialogue in Hell.*

As we have seen, Joly retrospectively claimed that as soon as the first idea for the *Dialogue* emerged he thought that Machiavelli 'would be Napoléon III, who would describe his own abominable politics'. In quoting this passage, later commentators failed to mention that it comes immediately after an allusion to Montesquieu *'embodying a whole side* of the ideas I wanted to express' ('comme *personifiant tout un côté* de mes idées que je voulais exprimer'). Montesquieu did not embody the entirety of Joly's ideas; but Machiavelli did not embody the entirety of the ideas and politics of Napoléon III either.

A passage from the ninth dialogue will prove this point. Machiavelli explains to Montesquieu that the new constitution issued after the *coup d'état* would be submitted to popular vote, to be either approved or rejected as a whole: an obvious allusion to the plebiscite which, on 2 December 1852, legitimized Louis Napoléon's claim to the emperorship – an unprecedented historical hybrid. Machiavelli quickly dismisses the American example: we are in Europe, he says; the idea of debating the constitution before voting would be absurd. A constitution must be the work of a single man, because 'things have never happened otherwise, as the histories of all the founders of empire testify – the Sesostrises, the Solons, the Lycurguses, the Charlemagnes, the Fredericks II, the Peters the First, for example':

> 'You are about to expound upon a chapter from one of your disciples,' remarks Montesquieu.
> 'Who?' asks Machiavelli.
> 'Joseph de Maistre' replies Montesquieu. 'Some general points you make are not without merit but I find them inapplicable here.'

Montesquieu is implicitly referring to a passage from Maistre's *Considérations sur la France*, chapter 6: 'On Divine Influence in Political Constitutions'. The passage reads as follows: 'No mere assembly of men can form a nation, and the very attempt exceeds in folly the most absurd and extravagant things that all the Bedlams of the world might put forth.'

To illustrate this scornful remark, Maistre quoted in a footnote a passage from Machiavelli's *Discourses on Livy* (I, 9): 'It is likewise essential that there should be but one person upon whose mind and method depends any similar process of organization.' Later, in the same chapter of his *Considérations*, Maistre ironically compared Montesquieu to a pedantic poet, and Lycurgus, the Spartan law-giver, to Homer. In other words, on the issue of constitutions Maistre invoked the authority of Machiavelli, not of Montesquieu, whom he regarded as a pedantic, ineffective theoretician.

Joly shared this attitude, since on the issue of constitutions he invoked the authority of Maistre, the arch-reactionary thinker, not of Montesquieu. I am referring to a passage from *Le barreau de Paris: Etudes politiques et littéraires* (The Bar of Paris: Studies in Politics and Literature): a series of generally satiric portraits of lawyers, sometimes disguised under pseudonyms, interspersed with general reflections, which Joly published the year before the *Dialogue in Hell*. In one of *Le barreau's* footnotes, Joly spoke contemptuously of 'the folly of constitutions and their inability to build anything', praised Maistre as 'a writer whose prophetic voice had, at the beginning of our century, an unchallenged authority', and quoted with approbation passages from Maistre's *Essai sur le principe générateur des constitutions et autres institutions humaines* (Essay on the generative principle of constitutions and other human institutions), including some that not only closely resembled the remark from the *Considérations sur la France* quoted above, but repeated the same reference to Machiavelli's *Discourses*.

Let me sum up my admittedly intricate argument. I compared four books, two by Maistre (*Considérations sur la France* and *Essai sur le pricipe générateur des constitutions*) and two by Joly (*Le barreau de Paris* and *Dialogue aux enfers entre Machiavel et Montesquieu*). The former book by Maistre is quoted in the latter; both of them are alluded to, either explicitly or implicitly, in Joly's two books, the composition of which partly overlapped (the reader of *Dialogue in Hell* will not miss a reference to 'machiavélisme infernal' ('infernal machiavellianism') in *Le barreau de Paris*). The four books can be regarded as pieces of evidence pertaining to the same context. But if we put them side by side, a greater, not a lesser, ambiguity emerges. The boundaries between reality and fiction blur: the fictitious Machiavelli develops arguments advanced by the real Maistre, who in turn developed arguments advanced by the real Machiavelli. Joly's emphatic praise of Maistre as a 'disciple' or a 'supporter' of the real Machiavelli (it is the Montesquieu of the *Dialogue* who says this) ultimately suggests a closeness to the fictitious Machiavelli as well. Joly seems to have projected something of himself into both characters of his *Dialogue in Hell*. If he shared Montesquieu's liberal orientation, he presented Machiavelli's arguments as stronger, not to say irrefutable: a painful split, which generated a dialogue based on a deep chasm between ideals and the perception of reality, wishes and thoughts – the very opposite of wishful thinking.

Joly's attitude towards the regime of Napoléon III was uncompromisingly hostile. But the *Dialogue in Hell between Machiavelli and Montesquieu* is more than a political pamphlet. Joly attacked Louis Napoléon's cynical abuse of power but also tried to make sense of a regime which he perceived as a new, original historical formation. For him, the plebiscite of 2 December 1852 meant far more than the *coup d'état* of 2 December 1851. Louis Napoléon's use of violence to crush his opponents was much less original than its outcome, in which police control cohabited with a free press, despotism and popular legitimacy. In order to understand those novelties one needed – so Joly suggested – the detached, unsentimental approach of an updated Machiavelli, not the illusions of a Montesquieu. But in Machiavelli's rueful assessment of the recent past there is no trace of the feeling of triumph we might expect from the alleged mouthpiece of Napoléon III. Joly's Machiavelli is a much more complex figure, in which the real Machiavelli (especially the author of *The Prince*), Napoléon III, and Joly himself overlap, creating a sort of composite portrait which recalls the photographic experiments Francis Galton began a few years later.

These blurred images may provide a visual equivalent of the ambiguity I am dealing with. In his effort to understand the Second Empire, Joly developed a complex, ambivalent relationship with the character who, under the name of Machiavelli, was meant to play the role of Napoléon III in the *Dialogue in Hell*. But the dialogic form also allowed the author a certain distance from his own creations. One can see Joly in the act of listening to himself as Montesquieu, aggressively challenged by himself as Machiavelli.

The voice of the fictitious Machiavelli is the voice of the enemy. I will refrain from recalling Carl Schmitt's over-quoted remark on the enemy (*hostis*) as the embodiment of our questions. I will instead cite a line from Ovid (*Metamorphoses* 4, 428) which Joly might have known: *Nam et fas est et ab hoste doceri* (learning is always legitimate, even from an enemy). Joly might have said: *especially* from one's enemy. From him one must learn the reasons for one's own defeat.

The modern form of despotism, Joly argued, includes free elections and freedom of the press. Clearly, he did not share the liberals' illusions about them; for him, real power lay elsewhere. When the *Dialogue in Hell* first appeared in 1864, this conclusion may have seemed paradoxical to many readers. It looks much less paradoxical today. Democracy is still, I believe, 'the worst of regimes, except for all the others', but its self-legitimization no longer convinces when, in the largest democratic country of all, only a minority of citizens cast a ballot in general elections – an event which is often, even for this active minority, the beginning and the end of their political commitment. Even more problematic is the voters' impact on the true centres of power and the decisions made there. At the beginning of the twenty-first century, democratic states are much more powerful than they were 150 years ago, when Joly wrote his analysis of modern despotism; their control of society is much more sophisticated; citizens are incomparably powerless.

All this throws some light on the twentieth-century reception of the *Dialogue in Hell*. In the 1920s and 1930s, as we will see, the discussion of Joly's book focused exclusively on its relationship with the *Protocols of the Sages of Zion*. Since the end of the Second World War, the *Dialogue* has been published three times in France, four times in Germany, twice in Spain, once in Argentina, once in Italy, and once, most recently, in the USA. The latest French edition, so far reprinted three times (1987, 1992, 1999), presented the *Dialogue* as 'a classic of politics which unveiled, one century before its time, the practices of modern despotism'. In a dense essay published as an introduction to that edition, Michel Bounan argues that Joly's analysis anticipated the world which has emerged from the collapse of totalitarian regimes – the world we live in. This conclusion, which Bounan has developed in other recent essays, reads Joly's work from the vantage point of the *Protocols of the Sages of Zion*, that legacy which no one, Joly least of all, could have foreseen. To clarify my position vis-à-vis Bounan's approach I will have first to say something about the relationship between those two works.

It has been said that, as a best-seller, the *Protocols* are second only to the Bible. This is probably an overstatement, but there is no doubt that every year new editions of the *Protocols* appear in the Middle East, in Latin America, in Japan, in Europe (I recall seeing the book prominently displayed in a bookstore in Budapest). As is well known, the *Protocols* claim to be the proceedings of a secret meeting of Jews (possibly hinting at the Zionist conference held in Basel in 1897) conspiring to infiltrate all sections of society: the economy, the press, the army and so forth. The ultimate outcome of this conspiracy will be a Jewish monarchy ruling the entire world. In some versions a 'translator's postscript' is appended to the *Protocols*, explaining that the book's contents represent a revival of a conspiratorial project initiated in 929 BC by Solomon and the Jewish elders.

An extensive scholarly literature has analysed in detail both the making and the incredible success of the *Protocols*. Let me sum up a few conclusions. As I mentioned at the beginning of this chapter, the first appearance of the *Protocols* was in Russia in 1903; other, somewhat different Russian versions came out in the following years. The worldwide diffusion of the *Protocols* began in 1919, with the publication of a German translation, hailed by *The Times* as an important document in the following year. Then, on 16, 17 and 18 August of 1921, Philip Graves, *The Times'* Istanbul correspondent, wrote a series of articles in which he claimed that the *Protocols* were a forgery; his proof was the many borrowings he traced from a forgotten book published over a half-century earlier: Joly's *Dialogue in Hell between Machiavelli and Montesquieu*. Graves had been informed of the connection between the two texts by an unnamed individual, later identified as Mikhail Raslovlev, a Russian emigré. Although some 'sources' of the *Protocols* had already been identified before, Graves' articles made a great impression. But the diffusion of the *Protocols* continued, stronger than ever, especially after 1933 (Adolf Hitler repeatedly stressed the authenticity of the *Protocols*). Mgr Jouin, the protonotary apostolic who had published the book in a French translation, had commented: 'Peu importe que les *Protocoles* soient authentiques; il suffit qu'ils soient vrais' ('It matters little whether the *Protocols* are authentic; it suffices that they are true'). One is reminded of medieval clerks fabricating *piae fraudes*, as they called them: forgeries inspired by the true religion. Some Catholics responded to the talk of forgery by declaring that Joly was Jewish, or a Freemason (he was neither), and that both the *Dialogue in Hell* and the *Protocols* could have derived from the same, still unknown, source. In 1934, two leaders of the Swiss Nazi party who distributed copies of the *Protocols* to promote the idea of a Jewish world conspiracy, were accused of defamation by some Jewish organizations. The judicial debate focused once again on the passages of Joly's *Dialogues* plagiarized in the *Protocols*.

But the word 'plagiarized' does not fully describe the relationship between the two texts. Certainly, the number of similar sentences is quite high. 'Like the god Vishnu, my press will have a hundred arms', says Joly's Machiavelli; 'like the Indian idol Wishnu, we'll have one hundred hands', say the Elders of Zion, in a chapter of the *Protocols* in which the Jews are urged to infiltrate journalistic organs of every political stripe. Parallel passages like this have been carefully identified. On a general level, there is a strong structural resemblance between the strategies ascribed to the Elders of Zion and to Joly's Machiavelli in their plans to control society: for instance, anti-Semitism will be turned to the advantage of the hidden Jewish forces, just as, by inciting political opposition groups, Napoléon III was supposed to consolidate power. How does one make sense of these resemblances?

The Truth About
"The Protocols"

A LITERARY FORGERY

From 𝕿𝖍𝖊 𝕿𝖎𝖒𝖊𝖘 *of*
August 16, 17, *and* 18, 1921

LONDON:
PRINTING HOUSE SQUARE, E.C.4.

FIGURE 2.3. The title page of the first article by Philip Graves exposing the *Protocols* as a forgery on the basis of a comparison with Joly's book.

There are two ways to answer this question; they supplement each other. The first deals with the historical connection between the two texts. Until recently there was a scholarly consensus, based on a series of textual clues, concerning the place and time in which the *Protocols* were made: France between 1894 and 1899. In a recent book, *Il manoscritto inesistente* (The Non-Existent Manuscript), Cesare G. De Michelis has advanced, also on the basis of internal elements, a different argument, suggesting that the *Protocols* were written in Russia in 1902–3. This hypothetical Russian origin is hardly compatible with the dependence of the *Protocols* on Joly's *Dialogue in Hell*: a nearly forgotten text, presumably unavailable in early twentieth-century Russia. De Michelis insists that the *Dialogue in Hell* was well known, but his reference to the Spanish translation published in Buenos Aires in 1898, thirty years after the latest edition, does not strengthen his case. De Michelis, who considers Joly's book a 'subtext' of the *Protocols*, relying on it to reconstruct the textual transmission of the *Protocols*, is ultimately compelled to assume, somewhat sketchily, that the (presumably Russian) authors of the forgery must have had some French connections, from whom they obtained either Joly's book or a series of quotations from it. This series of quotations would presumably have included other passages from French authors like Tarde or Chabry, also echoed in the *Protocols*.

We are back in France. But is it possible to identify a French link connecting Joly's book and the *Protocols*? Strangely enough, De Michelis does not mention an intriguing, albeit speculative, attempt to answer this question, though he refers to the book repeatedly: *L'Apocalypse de notre temps: les dessous de la propagande allemande d'après des documents inédits* (The Apocalypse of our Time: Unveiling German Propaganda on the Basis of Unpublished Documents). In this truly remarkable work, published in 1939 on the eve of the Second World War, and republished in 1991, Henri Rollin, a non-professional historian and member of the French secret service, explored in depth, with great insight and erudition, the context out of which the *Protocols* had emerged. Rollin discovered that in 1872 Joly began writing for *La liberté*, a right-wing newspaper. On the staff of *La liberté* was Edouard Drumont, a journalist who later became famous as the director of *La libre parole*, a newspaper confected of some Catholicism, a pinch of anarchism, and a large amount of virulent anti-Semitism. In his bestseller *La France juive* (1886), as well as in *Le testament d'un antisémite* (1891), Drumont mentioned Joly ('ce bon Jolly'), though he did not manage to spell his name correctly. In 1894, after the assassination of Sadi Carnot, the president of the French Republic, by an Italian anarchist, Drumont fled to Brussels, to avoid any fallout from his mildly pro-anarchist articles. In an interview published in the *Figaro* (18 July 1894), Drumont threatened to revive the aggressive pamphlets written under the Second Empire: 'we have to prepare some new *Propos de Labiénus*', he said. Then, pointing at a huge box, he shouted: 'Documents! Genuine documents! Until now I kept silent, driven either by compassion or by Christian charity. I fought a civilized war. But if we are to be outlawed by an unjust law, I will start a savage war.' At that moment, Rollin suggests, Drumont may have come across the *Dialogue in Hell*: the book written by his former colleague, which was more easily available in Brussels, where it had been published, than in Paris. Drumont's reference to 'some new *Propos de Labiénus*' is particularly intriguing: the work referred to was a satire against Napoléon III in the form of a fictitious dialogue between two ancient Romans, obviously inspired by Joly's *Dialogue* published one year before. On 10 January 1896, Drumont mentioned once again in *La libre parole* the possibility of writing a 'nice pamphlet' as a sequel to the *Propos de Labiénus*. Ten days later he came back to the same topic: 'If the *Dialogues of the Dead* were still in fashion . . .' All this does not prove that Drumont resurrected the *Dialogue in Hell* because of its anti-Semitic potential, presenting a fiction as a genuine document, nor does it suggest that he made that text available to the Russian makers of the *Protocols*. But, as Rollin implicitly suggested, this trail deserves to be followed. In 1898 – *l'année juive*, as Drumont ruefully labelled it – a series of dramatic events suddenly reopened the Dreyfus *affaire*. The document which allegedly proved Dreyfus' guilt turned out to be a forgery; Colonel Henry, jailed as the author of the forgery, committed suicide. Drumont took a defiant attitude. *La libre parole* launched a massive subscription in order to build a monument to Henry, a man (Drumont wrote) who had naively made a silly gesture, infinitely less relevant 'than the infamous means used by Jews to enrich themselves and become our masters'. On 26 February 1899, *La libre parole* published on its front page an article signed 'Gyp', pseudonym of Sibylle-Gabrielle Marie Antoinette comtesse de Mirabeau-Martel, well-known author of playful anti-Semitic articles. The article, entitled 'L'affaire chez les morts', was a grotesque variation on the genre 'dialogue of the dead' which had inspired Joly's *Dialogue in Hell*. Gyp presented Calvin, Joan of Arc, Catherine de Medici, Voltaire, Napoléon, Gavroche insulting and physically abusing Moses, Jeremiah, Meyer

Rothschild, Jacques de Reinach, all of them speaking French with a German accent. Today, some passages of this vulgar joke sound like an ominous prophecy. 'I have been so often criticized along history', Catherine de Medici says, 'but a St Bartholomew night against Jews would not surprise me at all.' The concoction of the *Protocols*, based upon Joly's *Dialogue in Hell* – a book nobody read any more – must have emerged in this context, and possibly in those months.

But the resemblance between Joly's *Dialogue in Hell* and the *Protocols* requires a different kind of approach, which impinges directly on the present. The *Dialogue* includes one – hostile – reference to the Jews, in a passage which surfaces again in the *Protocols* (except for the reference to the Jews, which was skipped). This circumscribed point of intersection is not very relevant. Much more interesting is the general isomorphism between the two texts, especially if we assume that Joly was able to decipher, in his analysis of Napoléon III's regime, the long-term phenomenon which he called 'modern despotism'. How should we interpret the *Protocols*, then? As a mere caricature? Michel Bounan suggested a different relationship: the *Protocols* were 'a police counterfeit of a revolutionary turmoil' ('la contrefaçon policière d'une agitation révolutionnaire'). This remark possibly implies August Bebel's famous definition of anti-Semitism as 'the idiots' socialism', but goes much further. According to Bounan, the real conspiracy which inspired the false one – the *Protocols* – is a classic example of a trait which defines the system described by Joly – that is, the modern state. This trait is defined as follows: '*a permanent hidden conspiracy* aimed at maintaining indefinitely the current state of dependency' (but the French text is much stronger – '*complot permanent occulte* de l'Etat moderne pour maintenir indéfiniment la servitude' – a word which echoes, perhaps unwillingly, Tocqueville).

I have few biographical data concerning Michel Bounan. As I guess from some of his writings, as well as from information available on the Internet, Bounan must have been close to Guy Debord and the *situationnistes*, the well-known far-left group which was particularly active during the May 1968 revolt in Paris. Today, Bounan seems to be the driving force behind a small publishing house which has recently republished two books I have been dealing with in this chapter, Joly's *Dialogue aux enfers* and Rollin's *L'Apocalypse de notre temps*. In a series of short, elegant essays, explicitly inspired by those works, Bounan has unfolded a consistent conspiratorial view of history. In modern societies, power is all-pervasive; false conspiracies and false aims divert the energies of everybody (with the exclusion of a small privileged élite); even the perception of being victims of injustice has been erased. Bounan's most recent pamphlet, *Logique du terrorisme* (Logic of Terrorism), published in 2003, approaches the events of the last few years from this perspective.

I have never agreed with the widespread attitude which automatically dismisses conspiracy theories as absurd. Of course, most often they *are* absurd, and sometimes much worse than that. But, as I noticed a long time ago, in a book devoted to the witches' sabbath stereotype, conspiracies exist, and false conspiracies always conceal real ones (a point also made by Bounan). After September 11, more people would agree that conspiracies exist; but I am well aware that an attempt to identify false conspiracies concealing real ones may lead to weird conclusions. Is it possible to trace a boundary between healthy scepticism vis-à-vis some official versions and conspiratorial obsession? Bounan, in my view, has overshot that boundary, inspired by the misleading principle *is fecit cui prodest*, which retroactively, and illogically, turns the achievement of an end into a causal relationship. (A government which exploits the political perspectives opened up by a terrorist by itself launching a military campaign is not necessarily responsible for the terrorist attack.) Bounan seems to have been mesmerized by his topic, the *Protocols* and their source, Joly's *Dialogue in Hell*. But to reject a conspiratorial view of history as an inverted version of the *Protocols* would be insufficient. In order to clarify my position I would like to go back to the relationship between Joly's *Dialogue* and the *Protocols*.

Joly paid a price for the brilliant literary form he chose for the expression of his ideas. His fictitious Machiavelli describes in detail, in the first person, the political strategies he would adopt to create a certain world – the world that in fact existed at the time Joly wrote. Joly's readers watched as this Machiavelli brought into existence through his own will the France they saw from their own doorways. Passing allusions to larger, anonymous phenomena, like the fragility of modern societies, were not developed. By imagining one man, one omnipotent individual who planned to shape society according to his intentions, Joly unwittingly paved the way for the unfortunate posterity of his work. The makers of the *Protocols* deliberately fitted materials taken from the *Dialogue in Hell* into a largely preexisting mould, developing a suggestion provided by

Joly himself. All ambiguities disappeared. A sophisticated political parable was turned into a crude forgery.

FURTHER READING

Cohn, N., *Warrant for Genocide. The Myth of the Jewish World Conspiracy and the Protocols of the Elders of Zion*, London: Eyre and Spottiswoode, 1967.

De Michelis, C. G., *The Non-Existent Manuscript. A Study of the Protocols of the Sages of Zion*, trans. R. Newhouse, Lincoln, NE: University of Nebraska Press, 2004.

Joly, M., *Dialogue aux enfers entre Machiavel et Montesquieu*, ed. Michel Bounan, Paris: Allia, 1999 (Joly, M., *The Dialogue in Hell between Machiavelli and Montesquieu*, ed. J. S. Waggoner, Lanham, MD: Lexington Books, 2003).

Rollin, H., *L'apocalypse de notre temps: les dessous de la propagande allemande d'après des documents inédits*, Paris: Gallimard, 1939 (1991).

Taguieff, P.-A. (ed.), *Les Protocoles des sages de Sion*, 2 vols., Paris: International Berg, 1992.

3 Evidence for Catastrophes in the Evolution of Life and Earth

VINCENT COURTILLOT

Introduction

The *Oxford English Dictionary* offers three main definitions of the word 'evidence'. The first meaning of evidence is simply 'clearness', the very fact of 'being (self?) evident'. The second meaning, more cautious, is perhaps closer to that which most people might expect, 'grounds for belief', a 'testimony of facts tending to prove or disprove something'. It is interesting to see the words 'belief' and 'prove' side by side and we may have to return to the relationship between these two words in scientific enquiry. Thirdly, we step down to the meaning known to most television watchers, that is 'information given in legal investigation to establish a fact'. We are going to embark on a short voyage, where our question, or sought-after goal, will be 'is there any evidence for catastrophes in the history of Earth and in the history of Life on Earth?' It is an engaging question for many reasons, not least because it brings together different scientific disciplines, each with differing methods and expectations. Put simply, up until recently the history of Earth has been the sole province of geologists, and the history of Life largely that of palaeontologists. These two disciplines now pay more attention to each other (at least sometimes) and have been joined in a very interesting way by biology, which has access to the distant past in a manner which no one would have thought of until a few decades ago: the study of genetics. In our quest to identify and understand catastrophes we will be using most definitions of 'evidence', often implicitly, and the reader should be cautious and question precisely which meaning we are implying on each occurrence of the word. We will start with the third meaning, asking whether geologists and palaeontologists have provided us with prima facie evidence, in the form of direct testimony, tending to prove that such catastrophes have indeed occurred. Therefore, here the scientific investigation will be a bit like a legal investigation, in which we will try to collect and weigh evidence for these catastrophes of the past.

The disciplines of geology, palaeontology and biology as we recognize them today have been constructed over the last two and a half centuries, in the cause of which the precise questions we are asking relating to catastrophes have seen much refinement, thanks to the work of hundreds of scientists, in the last few decades. The debate about whether or not Earth's history has been punctuated by catastrophes has existed from the earliest days of modern science and is told in many excellent books and textbooks. When geologists and others speak of catastrophes, they mean an event which itself is short-lived and yet has a vast amplitude. Of course, our most immediate problem is that that which seems short-lived on a geological scale may nonetheless have a very long absolute duration and the reader will have to become familiar with the profoundness of geological time (in the sense of the historian Fernand Braudel). The measuring clock of the geologist uses a million years where we would use a week, in that a man who lives to be a 100 has seen approximately 4800 weeks, whereas Earth (together with the Sun and other planets) was formed 4567 million years ago. With that simple conversion rule in mind, the geologist's 100 000 years is like a day in our lives, and a year is to Earth what a second is to us. Geologists divide the history of Earth into three consecutive periods, familiar from dry school lessons: Palaeozoic, Mesozoic, Cenozoic. The first true mammals and dinosaurs belong to the Mesozoic. If we were to reckon the entire history of Earth as a year, then the Mesozoic period covers only two weeks of the last month, from 11–26 December, when the Cenozoic begins. The human race emerges at 2 p.m. on New Year's Eve. We must become blithe in our treatment of these mind-boggling expanses of time!

And so what about the debate on catastrophes? Put simply, on the whole most scientists have come down in favour of the principle known as uniformitarianism, which suggests that we should never invoke as an explanation for our evidence a catastrophe, or indeed any event or process, not compatible with our current knowledge of the laws of Nature (physical and chemical). This seems very fair, but there is a more dogmatic version of uniformitarianism, which has been the majority view in the Earth sciences until recently, which insists that not only the quality but also the quantity of Nature's laws has not changed. In other words, one should not invoke a phenomenon in the distant past with an amplitude much larger than that which Earth has recently witnessed (recently having to be defined in the context of geological time). We will argue that this version of uniformitarianism is wrong and that catastrophes, which is to say events with amplitudes hugely greater than has recently been the case, have profoundly changed the course of biological evolution and the geological history of Earth. These catastrophes had durations ranging from a few years (or even less) to hundreds of thousands of years.

The debate is certainly not new; it has been gaining momentum since the decades around the year 1800. This was an exciting period for science, in which geologists and palaeontologists were beginning to ask, in ways which seem to us to be strikingly up-to-date, where our origins are and where the origins of the Earth are. Two schools of thought emerged. At the end of the eighteenth century, the famous French naturalist Georges Cuvier pointed to the sudden disappearance of certain fossil species at a given level in the rocky strata of the Paris Basin and the subsequent emergence of new species which did not seem to have been there before. He believed that this provided evidence that Earth's history had been punctuated by many extinction events, during which entire faunas were eliminated by large-scale catastrophes which he did not specify precisely (although he looked to a divine explanation and suspected floods of a biblical sort). However, the British geologist James Lyell negated catastrophism in his *Principles of Geology* (1830) on the grounds of dogmatic, or substantive, uniformitarianism. His ideas were profoundly influential, paving the way for many branches of modern geology, and in his 1859 masterpiece on the evolution of species, Charles Darwin, a friend and ardent supporter of Lyell, paid very little attention to the extinction of species. Darwin, Lyell and others considered extinction to be a natural part of the evolution of Life, albeit sometimes seeming intense and rapid in the geological record for lack of proper archives (i.e. missing rock strata) and dating methods (the apparent amplitude being due to the enormous lengths of geological time). In the midst of this debate, in 1830, the geologist William Whewell adopted a much more moderate position, which it is hard not to see as both brave and 'modern'. He defended the rather less dogmatic strain of uniformitarianism and argued that it was very bold indeed of Mankind to assume it had been around long enough to sample the full range of amplitudes that the Earth could actually produce. It could very well be the case that circumstances far outside the experience of Man have shaped the world. In essence this is the argument I will present in the following pages and, if everyone had been convinced by Whewell

174 years ago, perhaps I would not need to do so. (It is worth noting that, as with many great thinkers, the followers of Lyell and Darwin have often somewhat overstated their mentor's position.)

Let us now turn to the evidence we think we have, pertaining to these questions. Geology and palaeobiology are historical sciences, and in this sense are a little different from physics and chemistry. Of course, the notion of time is there in physics and chemistry, but the notion of deep time brings its own difficulties, and the evidential problems encountered by geologists and palaeontologists are in some ways like those encountered by the historian: one can be forced to work with very imperfect material and oblique evidence. The evidence can come in various strains: it can come from direct *observation* of rocks or living or fossil beings in the field; it can arise from *experiments* performed in the laboratory or from *measurements* made in the laboratory on specimens brought back from the field; it can come from *numerical simulations* which are gathering increasing importance (not without some dangers) due to the remarkable advances in computing technology; finally it can be considered as the result of *reasoning*, by logic and inference, or *model building and testing* based as much as possible on the other sorts of evidence listed above.

Evidence from palaeontology

So let us proceed with a brief summary of our evidence, beginning with observations from fossils amassed by palaeontologists over the last 200 years and more. The work of generations of palaeontologists in identifying and dating fossils, and placing them in their proper relationship to each other, has allowed us to see that the evolution of Life on Earth has not been a quiet process. These palaeontologists are far from unanimous about the nature and duration of the great upheavals which have disquieted Earth, and evidence from palaeontology can be somewhat ambiguous. So, for instance, if a fossilized bone is discovered which is believed to be the last from a particular species it certainly need not mean that it is from the very last animal of that species to have lived; the species may have prospered for millennia afterwards without any example being fossilized. Fossilization is a rare event to start with, and most fossils that do form have not been and will not be found. Some that are found may have been eroded by a stream and redeposited in a much younger terrain leading to grave difficulties in dating. The complications are many, but it is possible to make some fairly certain assertions about the diversity of Life on Earth over the last 600 million years or so.

The bottom of Figure 3.1 shows the increase in the number of families of extinct fossil species in the record of marine sediments, implying a marine environment. This is expected to be an indication (or proxy) of changes in the diversity of Life on Earth (or biodiversity) since the time when we have sufficient archives, the beginning of the Cambrian era, the first stage of the Palaeozoic era, which began some 550 million years ago. In order to be more rigorous, we should have indicated uncertainties on the individual values shown on the figure; neither exact time nor exact number of families are known and there should be a small 'cloud of uncertainty' around each point. But the main features of the curve are believed to be reasonably safe (scientists say 'statistically robust'): an exponential rise in diversity, the so-called Cambrian explosion, followed by a plateau with rather constant diversity for almost 200 million years. The plateau does not mean that there were no extinctions, simply that extinctions and the appearance of new species were approximately balancing each other out. Some notches in that plateau, for instance at 400 and 360 million years ago, are believed to be meaningful, but our first shock comes at 250 million years ago (a little less in the figure which has somewhat outdated age estimates) when half of the families of species disappear in what seems like a brief geological event. Because of the hierarchical nature of biological classification, 50 per cent of families means over 95 per cent of species. The remaining 5 per cent of species are likely to have been hard hit, even though not rendered extinct, and possibly more than 99 per cent of all then-living individuals perished. This event has been known of since the nineteenth century and was actually used to define the separation in the stratigraphic record between the Palaeozoic and Mesozoic eras. It is the worst crisis suffered by Life on Earth that we know of; we were within less than 1 per cent of seeing Life on Earth extinguished. Its actual duration has been made more precise by careful dating (mostly using radiogenic isotopes of potassium and uranium, the long-term geological equivalents of the carbon 14 clock used, for instance, by archaeologists) and is known to have been of the order of only 1 million years, an amazingly short time by geological standards. After that event, diversity rose again, to suffer a new smaller blow at the end of the Triassicera, 200 million years ago, then rose again, then suffered its most recent major drop 65 million years ago. This mass extinction is used to mark the end of the Mesozoic era, and the beginning of the

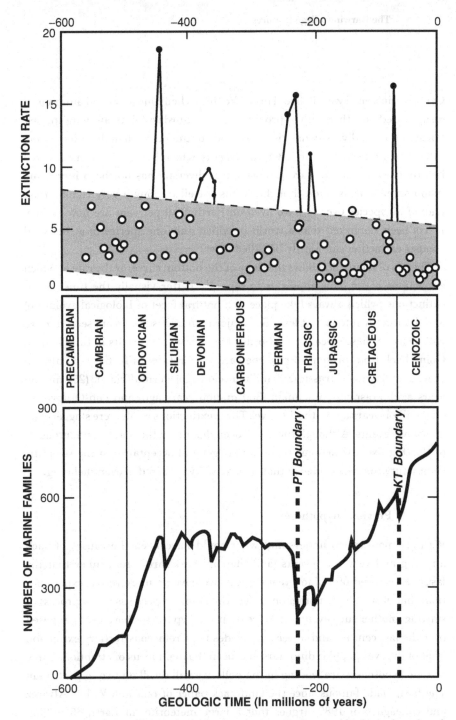

FIGURE 3.1. Changes in species diversity (actually illustrated by the total number of marine families rather than species) (bottom) and extinction rate (measured as number of families becoming extinct per million years) (top) as a function of time (after David Raup and John Sepkosti).

Cenozoic, in which we still live. This is the time when dinosaurs and ammonites disappeared, together with approximately two-thirds of all then-living species. Since that time, diversity has risen fast and continuously, though at decreasing rates, and we live in a world with a diversity which has never been exceeded before. However, as we have just seen, this diversity has not been increasing monotonously, as is often thought. It has travelled along a very chaotic road and, of all species ever to have lived on Earth, 99.9 per cent are now extinct. It has been remarked that a world in which only one species in a thousand escapes extinction can hardly be called safe!

The top of Figure 3.1 shows the slope of the bottom curve, or the rate at which extinctions occur. This allows us to separate better visually the times when extinctions (which always take place as a natural facet of biological evolution) are abnormally intense. At least five major extinctions events are seen, and we will focus on these so-called first-order mass extinction events in this chapter. Going back from present to past they are the Cretaceous-Tertiary (65 million years ago), Triassic-Jurassic (200 million years ago), Permo-Triassic (250 million years ago), Frasnian-Famennian (360 million years ago) and end-Ordovician (430 million years ago) – the 'big five'. These extinctions really were staggeringly enormous events. Although there has been debate on the reality, and intensity, of the 'big five' extinctions, there is now general acceptance in the scientific community that the evidence that they actually occurred is overwhelming.

Previous hypotheses

We must now try to understand what could have been the causes of such amazing and very rare events (and they are very rare – remember that the last one occurred 65 million years ago, so we have no more recent experience than that of such a phenomenon, unless the Human species is in the process of starting another such event . . .). More than 100 hypotheses have been proposed over the last century and in recent decades the Cretaceous-Tertiary extinction (that of the ever popular dinosaurs) has been the main focus of attention. Three or four hypotheses are leading in the polls: we will dwell on two in particular. The first, made famous since the landmark paper of Luis and Walter Alvarez and colleagues in 1980, argues that a large meteorite hit Earth 65 million years ago, injecting huge amounts of dust into the atmosphere and leading to a severe, generalized ('nuclear') winter. This would have been accompanied by huge wildfires, acid rain, and a greenhouse climate due to the injection of large masses of carbon dioxide, a greenhouse gas, into the atmosphere. The second, which can be traced back to a number of authors and gained momentum with the work of my team in India, proposes that a huge volcanic field erupted at much the same time as the meteorite impact. Volcanic action may explain the nature and extent of the extinction event more completely. With each massive eruption, sulphur dioxide injection would have led to cooling, carbon dioxide to subsequent warming, with catastrophic consequences for Life. The large number of superimposed layers identified at the Deccan traps, a stack of flows of balsatic lavas across much of northeastern India, indicates hundreds of eruptions in a relatively short period of time, and the process of atmospheric cooling and heating would have attended each one. There are two less publicized yet important hypotheses. One relates to slow change in sea level and alteration of environments conducive to more or less continental climates, and the other proposes that complex ('non-linear' in mathematical terms) interactions between living species can generate catastrophic extinction events without recourse to external forcing factors.

The media often like to pit scientists against each other, thinking that combat will attract a greater audience. Debate is certainly healthy, but war may not be necessary. Generally, when I have had to explain the 'volcanic hypothesis' in public, in the media, it has been expected that I would reject the 'impact hypothesis'. But the evidence for an impact at the Cretaceous-Tertiary boundary, or KT boundary as it is commonly known (since the nineteenth century, geologists across the world have used K, from the German word *Kreide*, meaning chalk, to signify the Cretaceous, and T for Tertiary), is excellent. There are many aspects to that evidence, and I will only mention the two (in my view) strongest. In all well-preserved stratigraphic geological sections, and well over 100 have now been described, the KT boundary is marked by a thin (1-cm thick most of the time) layer of clayey material that is unusually enriched in the metal iridium. Iridium is exceedingly rare in the Earth's crust, but is more enriched in meteorites, and probably also in Earth's deeper layers. Also, the clayey layer sometimes contains tiny grains of quartz, which display unusual features when observed under a microscope. These lamellae, as they are known, form only when a strong shock wave passes through the crystal, such as might have been caused by a meteorite impact. Moreover, geophysicists looking for oil and analysing the gravity field of the surface of the Earth in the northern part of Mexico's Yucatan Peninsula have found what is very likely to be the impact crater, buried under thousands of metres of more recent sediments. The crater, some 150 km in diameter and hence the largest preserved on Earth, has been dated precisely at 65 million years. There is little doubt in my mind that this event did happen at the time that the iridium-bearing layer was formed, which is now taken by definition to be the KT boundary. It must be said that some difficulties and contradictions have been identified recently. An analysis of sediments from the cores in and around the crater has led the Princeton-based palaeontologist Gerta Keller to believe that the Mexican impact actually happened a few hundred thousand years before the KT boundary itself. This is a topic of heated ongoing dispute.

Palaeomagnetism and the Deccan traps

For all that it is often taken to be compelling, there is a problem with the meteorite hypothesis in accounting for all lines of field evidence. Many palaeontologists believe that they see several events of mass extinction in the detailed fossil record prior to the iridium level (Figure 3.2). These 'stepwise' extinctions would need a sequence of events occurring over a geologically brief, but not instantaneous, period of a few hundred thousand years. In other words, a catastrophic impact may not be the whole story and this is where the volcanic hypothesis comes in. The story is one of serendipity, and the arrival of myself and my colleagues in the debate was an entirely unexpected event. In fact, our involvement seems to indicate what has often proved to be the case before, that scientific inquiry does not always proceed in a logical Cartesian manner from problem to solution, but rather sometimes requires a new start without preconceived ideas, almost a return to ignorance. I have told the story in a book (Courtillot, 1999) which I will attempt to summarize here.

My area of research is geophysics, and my particular speciality is palaeomagnetism. Many rocks freeze the direction of the Earth's magnetic field at the time when they were formed. What my colleagues and I do is gather rock samples from geological environments of (hopefully) well-known age and structure, bring them back to the laboratory and measure their fossil magnetization. If one can retrieve that direction, its orientation in the horizontal plane tells you in which direction north pointed when the rock was formed, and its plunge under the horizontal tells you at which latitude it was formed. This is because

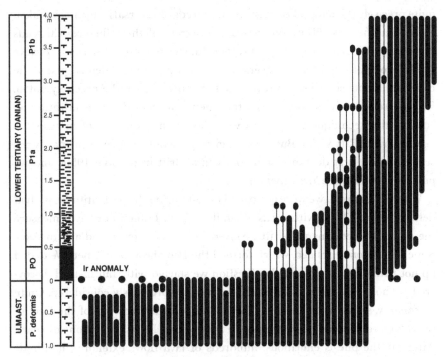

FIGURE 3.2. Species distribution in the El Kef (Tunisia) section (after Gerta Keller). Each vertical bar represents the presence of a given species. Geological stages, paleontologic zones and sedimentary facies are given as columns to the left. The reference Cretaceous-Tertiary (KT) boundary, defined by the spike in iridium concentration, is indicated.

Earth's natural magnetic field has the simple geometry of a bar magnet, or dipole, aligned with the rotation axis: the field is vertical at the poles and horizontal at the Equator and takes all intermediate positions in between. This was first described in Latin in 1600 by William Gilbert, private physician to Elizabeth I and a physicist in his spare time, in his treatise on the magnet, *De Magnete*. For our purposes the next major breakthrough was made around 1904 by a Frenchman, Bernard Bruhnes, who in the course of measuring the magnetization of lava from the Massif Central found that some was magnetized in exactly the opposite direction to that of the present-day field. Further investigation suggested that the Earth's field flips every now and then, pole to pole. The last time was 780 000 years ago, when the south and north magnetic poles were exchanged. This theory has been confirmed, and we now know quite precisely the dates of the last few hundreds of reversals which have taken place over the last 150 million years. A diagram of these (Figure 3.3) reads like a powerful 'bar-code', and is a wonderful chronometer. If one can retrieve the bar-code of sequences of reversals in a rock pile, and match it against the complete sequence of known reversals, then that pile can be precisely dated. Because rocks such as lava flows retain the memory of the magnetization on cooling, this technique has helped with dating the evolution of man in the Great African Rift Valley. But for geologists, palaeomagnetism is primarily the science which first demonstrated continental drift in the mid-1950s and we palaeomagnetists are 'drift measurers'.

With this in mind, we were interested in comparing the past latitudes we had determined for Tibet with rocks of similar age in India, in order to measure the amount of continental drift involved in the collision of India with Asia, some 50 million years ago, which formed the Himalayas and Tibet. We ended up not being able to answer the question we started with, but instead discovered a whole new series of questions we had not previously considered. Quite reasonably, scientists in the early 1980s believed the thick pile of the Deccan traps (Figure 3.4) had been erupted in the course of tens of millions of years. After all, the pile is enormous: hundreds of lava flows extending hundreds of miles, each one comprising hundreds of cubic kilometres. We (and several other groups, mostly in India) measured the magnetization of the hundreds of basalt flows and found that only two polarity reversals had been recorded in the pile! This signified that the pile had been formed in an incredibly short time . . . but when? We found the fossil of a tooth of a ray fish preserved in sediments deposited by lakes that had formed in between two eruptions (and are therefore now sandwiched between two lava flows). This particular fish is known only from the very end of the Cretaceous period, and that, together with potassium–argon absolute dating and the fact that the time gap between the two magnetic reversals 'frozen' in the lava flows matched only one place on our bar-code of reversals, allowed us to show back in 1986 that the whole lava pile must have been erupted in only a few hundred thousand years at the time of the KT boundary, 65 million years ago! More recently, Narendra Bhandari and his colleagues from the University of Ahmedabad in India have shown that some lake sediments found in between lava flows in the northwestern part of the Deccan have preserved a blurred record of the impact in the form of anomalous concentrations of iridium; this has been confirmed by my colleague

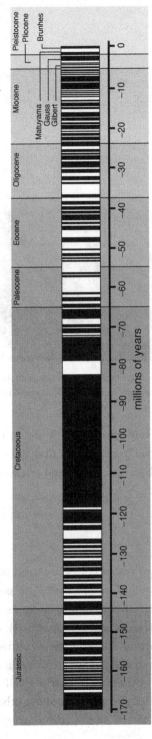

FIGURE 3.3. The magnetic reversal time scale, since the Jurassic era, 170 Ma ago. Normal periods are in black, reversed ones in white. The names of some magnetic 'chrons' (chronozones) are given (we are still in the Brunhes).

FIGURE 3.4. The Deccan traps in the western Ghats, Mahabaleshwar, India, showing their staircase-like erosion profiles (photograph by the author).

Robert Rocchia. This is very important since it shows that the impact and the volcanism coincided in time. It also shows that the volcanism had begun before the meteorite struck Mexico, and so could not have been triggered by the impact as some had previously suggested.

How many impacts?

We now have evidence that two catastrophes occurred on Earth at the same time, though with vastly different durations, yet both geologically short. For me the question was no longer whether one hypothesis was right or wrong. Instead two questions immediately came to mind: (1) what part did each catastrophe take in the extinction scenario (the eruptive episodes matched nicely the stepwise extinction indicated by the stepwise disappearance of fossils in the stratigraphic record, whilst the impact matched well the iridium layer and a major mass extinction episode), and (2) what about other mass extinctions? Scientists do not like to base a theory on a single occurrence. It was time to go and look in more detail at other potential cases of impacts and trap eruptions and to see whether any coincided in time with other mass extinctions. As far as impacts are concerned, there are none remaining at the surface of Earth which are as large as the KT Mexican one. Of course such craters may have existed and been erased by erosion, or covered by more recent sediments, or await discovery in the oceans, or even may have been subducted, that is returned by plate tectonics to the mantle at depth under active volcanic arcs such as the Pacific rim of fire. There are large craters, with diameters of the order of 80 km, which are now reasonably well dated and do not coincide with the time of an extinction. This implies that the impact of a meteorite somewhat smaller than the KT one, yet nonetheless able to excavate an 80-km-diameter crater and most likely destroy any life in a wide area surrounding the impact, does not reach the intensity level that translates as a mass extinction event such as might appear in the fossil record. We must realize that, however horrendous, such a large impact is a minor event in the evolution of Life on Earth, orders of magnitude smaller than what is needed to wipe out more than half of all species on Earth. There have been several claims in the scientific literature over the last decade for indirect evidence of impact in the form of unusual chemicals (such as fullerenes, or carbon 'buckyballs' with exotic ratios of certain isotopes of argon, for instance) but none has stood the test of critical analysis by other laboratories. In my view, the KT impact remains the only documented example of an impact coinciding with a mass extinction in the past few hundreds of millions of years.

How many traps?

So, let us now turn to occurrences of flood basalts such as the Deccan. There are indeed others, though not that many. Figure 3.5 shows most known occurrences of continental (subaerial) and underwater (oceanic) plateau basalts. There are about a dozen that were erupted in the past 300 million years; some, including the Siberian, Ontong Java, Caribbean and Parana traps, are truly massive. Although they are not all in the same state of preservation, they share many characteristics. When reconstructed in one's mind eye, correcting for subsequent erosion, they appear to have had original volumes in the order of 1 to 10 million cubic kilometres, thicknesses of a few kilometres and horizontal coverage in the order of hundreds of thousands of square kilometres.

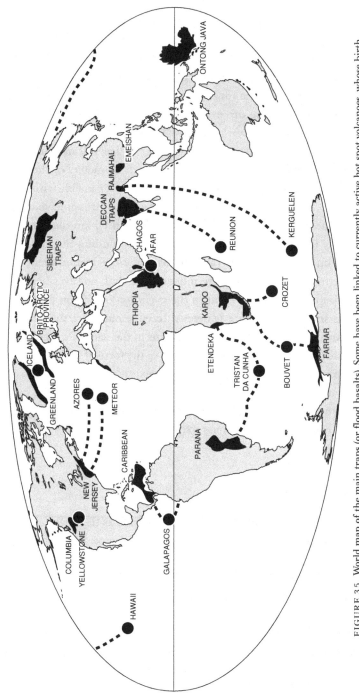

FIGURE 3.5. World map of the main traps (or flood basalts). Some have been linked to currently active hot spot volcanoes, whose birth may be the cause of the traps.

That is something like covering the entire British Isles with lava to the depth of seven or eight kilometres. Each seems to have been erupted in not much more than a million years, again a rather short time for such a volume from a geological standpoint. But more importantly, their ages, which are becoming better known through the concerted efforts of several geochronology laboratories, appear often to coincide with mass extinction events. For instance, Figure 3.6 shows a reconstruction of Earth 200 million years ago, at the time of the Triassic-Jurassic mass extinction event. Well this is precisely the time of eruption of the huge so-called Central Atlantic Magmatic Province (CAMP), which covered more than 9 million square kilometres, and erupted over the area that was to rift shortly thereafter and to become the Central Atlantic Ocean. In the course of the extinction associated with this eruption, almost all ammonites disappeared, together with over half the bivalve families, many brachiopods, gastropods, corals and sponges and 80 *per cent* of quadrupeds. Figure 3.7 takes us 50 million years earlier, when the large Siberian traps erupted, precisely at the time of the Permian-Triassic mass extinction. The Permian-Triassic is the mother of all mass extinctions, during which Life was almost extinguished, and the Siberian traps indicate an extreme period of volcanism. Situated on the north-western margin of the Siberian platform, in places these flows have a cumulative thickness of almost 4 km even today, and before erosion their volume must once have been in excess of 2 million cubic kilometres. I will not review all trap events, but I would like to note that when one reconstructs Wegener's Pangaea supercontinent prior to its break-up, we see that three traps erupted in succession at the very locations where ocean basins were to open shortly thereafter. The 200-million-year eruption of the CAMP was followed by the opening of the Central Atlantic as we have just seen; the 135-million-year eruption of the Parana traps (in South America, with a small piece, the Etendeka traps now lying in Namibia on the African side) was followed by the opening of the South Atlantic; and the Greenland traps (also called the North Atlantic Tertiary Volcanic Province or NATVP) were erupted 60 million years ago, signalling the opening of the North Atlantic basin. Following eruption of the traps, the hot source in the mantle which had produced them lingered as a 'hot spot', burning a series of what are now underwater extinct volcanoes, leading to the still active islands of Iceland (for the NATVP), Tristan da Cunha (for the Parana-Etendeka) or Reunion in the Indian Ocean to come back to the Deccan traps. It therefore seems that the present-day geography of our major

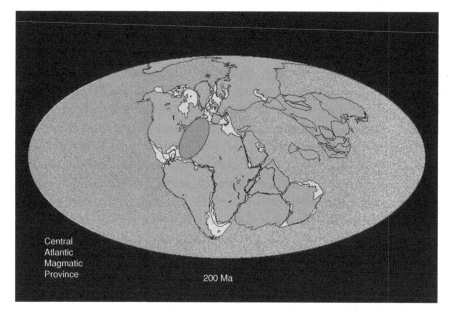

FIGURE 3.6. A reconstruction of Earth 200 million years ago, at the time of the Triassic-Jurassic mass extinction event, showing the enormous Central Atlantic Magmatic Province (CAMP), which covered more than 9 million square kilometres, and erupted over the area that was to rift soon afterwards to become the Central Atlantic Ocean (courtesy of Jean Besse).

ocean basins (at least those that have not been affected by subsequent convergence and subduction, as is the case for the Pacific) is inherited from the arrival (another kind of 'impact') at the Earth's surface of huge volcanic pulses.

Any scientific theory worth its salt must be able to stand as a basis for prediction and accommodate new evidence. In recent years palaeontologists have discovered that the wave of extinctions at the end of the Palaeozoic actually could be resolved into a pair of short, catastrophic events, the one we have seen at 250 million years linked to the Siberian traps, and another one at 258 million years, the end of the so-called Guadalupian stage of the Permian. So, we 'predicted' that there should have been a trap at 258 million years, and indeed one has been found since, in southwestern China, in the Emeishan region, which has been dated at the expected age, with an uncertainty of 2 million years. And even more recently, working with our Russian colleague Vadim Kravchinski, we may have found the trap that would be linked to the

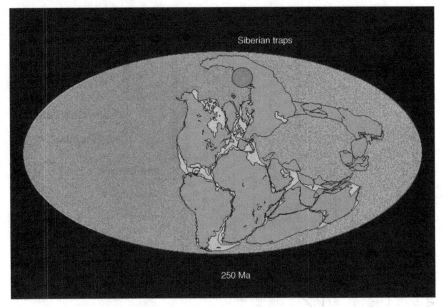

FIGURE 3.7. A reconstruction of Earth 250 million years ago, showing the large Siberian traps. The Siberian traps erupted at precisely the time of the Permian-Triassic mass extinction during which Life on Earth was almost extinguished. The Siberian traps indicate that this was a period of extreme volcanism (courtesy of Jean Besse).

previous mass extinction event at 360 million years ago, though at the time of writing this is still being dated in two different laboratories.

Is a correlation strong enough a basis for evidence?

The correlation between ages of flood basalts and ages of mass extinctions is shown in Figure 3.8. This is the third revision since we first proposed it in the early 1990s. As time has passed, the correlation has steadily improved, with more and more points on the correlation line and smaller age uncertainties. It is now clear that the last four, and possibly five, largest mass extinctions coincide in time with an identified trap without one miss. Other traps tend to correlate with smaller extinction events, or events recorded in the oceans as 'anoxic events', times when the bottom of the world ocean was deprived of oxygen. All in all, of the twelve principal balsatic regions, ten have been matched with

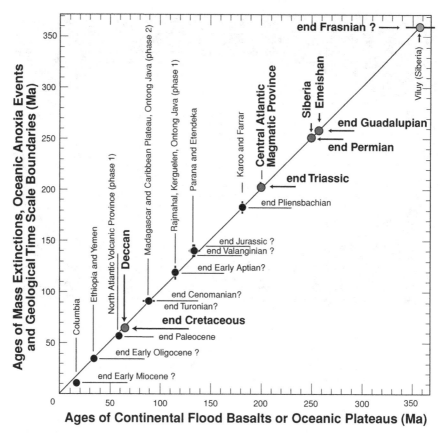

FIGURE 3.8. The figure shows correlation between ages of flood basalts and ages of mass extinctions. It is now clear that the last four, and possibly five, largest mass extinctions coincide in time with an identified trap without one miss. Other traps tend to correlate with smaller extinction events (Courtillot and Renne, 2003, 113–40).

periods of catastrophic disruption to Life on Earth. There is less than 1 chance in 100 that this could be a random, coincidental sequence. In contrast, only one impact coincidence has been found in the same time interval. It is therefore now rather safe in my view to propose that the eruption of flood basalts is the principal mechanism which triggers mass extinction events when eruptions occur directly into the atmosphere (that is, in most cases) or anoxia events when the eruption is submarine (which occurred a few times). A number of impacts have occurred which did not generate a sufficient biotic catastrophe to be recorded as a mass extinction event. Only the Mexican impact does coincide with a mass extinction, that at 65 million years. But this happened when the world was already under the spell of a few hundred thousand years of Deccan volcanism. It is therefore reasonable to propose that the extinction coinciding precisely in time with an impact was greatly amplified because the environment was already under severe stress and had already undergone the first steps of mass extinction. Had the meteorite fallen on Earth at a time of no extreme volcanism, it might not have left a trace in the palaeontological record (or it might, having been the largest meteorite to hit Earth in 300 million years; this is still an open question).

Attempting to model the climatic consequences

Now this coincidence is significant evidence but not final proof. It suggests a causal connection, but that connection may be rather complex. So the next step is to attempt to reconstruct the actual chain of physical events by which extreme volcanism could have led to the death of so many species and count-less individuals. This is the direction that our research efforts have taken most recently. Volcanic eruptions in the modern world, for instance at El Chichon or Pinatubo, have shown that injection of large masses of sulphur into the atmosphere could lead to significant climate change, in the form of large-scale (not necessarily global) cooling, by tenths of a degree Centigrade. The link between volcanic eruptions and climate change was first suggested by Mourge de Montredon in Montpellier (France) in 1783. De Montredon noticed a dry, sulphur-rich fog, and an unusually warm summer, followed by a particularly harsh winter. He highlighted disease of vegetation in that year. The fog was reported as far away as China and North America. When de Montredon heard, from navigators travelling from Iceland to Norway, that Iceland had seen much volcanic activity that year, he wrote to the Academy of Sciences in Montpellier proposing that the two events could be linked and that the fog could in part be the result of the eruption. Benjamin Franklin independently made similar suggestions, pointing out that volcanic emissions could limit the amount of sunlight reaching Earth's surface. Latterly, the Icelandic eruptive sequence has been reconstructed with great care by the volcanologists Steve Self and Thor-valdur Thordarson. It centred on the Mt Laki area in the southeastern part of the island, began in early June 1783 and continued with about ten paroxystic, explosive events, accompanied by quieter effusive episodes, which lasted until February of 1784. In the course of these eight months, 15 cubic kilometres of lava were emitted, along with large masses of tephra (rock particles) and sulphuric aerosols (a sulphuric mist). These were injected high into the stratosphere, up to 13 km high, according to eyewitness reports. This was the largest volcanic eruption in historic times. The gases emitted caused Iceland's greatest ever famine, which led to the death of more than half of all livestock and a quarter of the human population. More terribly, but also more interestingly for our purposes, the eruption seems to have had widespread consequences beyond Iceland. John Gratton and colleagues have analysed burial rates in England from 1770 to 1800. They find (Figure 3.9) that there is a distinct peak in the year of the Laki eruption. They have continued that work in more detail looking at monthly burials in various parts of France and come up with clear excess deaths in the months that followed the onset of the eruption. It does seem that the eruption, or rather the cloud of sulphuric aerosols it emitted which was subsequently dispersed over much of the Northern Hemisphere, was responsible for an unprecedented peak of deaths. Perhaps as many as 100 000 more people than was usual died across Europe in those few months, almost certainly through breathing-related problems brought on by increased toxicity. This is something like 20 times more than died as a result of the heat wave in the summer of 2003.

The Laki eruption has provided a starting point for a serious consideration of the processes and consequences of massive volcanic pollution. For the purposes of climate study, the atmosphere is divided into about twenty layers, which are themselves divided into small squares, several hundred kilometres on each side, known simply as 'cubes' or 'volume elements'. We now have at our disposal powerful numerical codes that allow us to calculate the consequences for climate globally of injecting pollutants into a given cube or cubes, allowing for known variables of temperature, pressure, winds, etc. The smaller the volume element the better the accuracy (or rather spatial resolution) of the model. My colleagues, Frédéric Fluteau and Anne-Lise Chenet and I have used the model built by the Laboratoire de Météorologie Dynamique in Paris and the volcanic history reconstructed by Thordarsson and Self to attempt to simulate the climatic changes forced by the Laki eruption. The result is quite promising. For instance, we are able to model reasonably well the geographical shape and

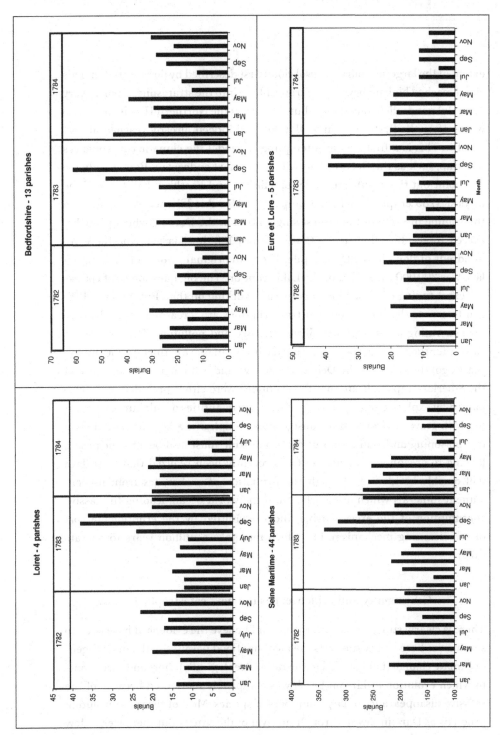

FIGURE 3.9. An analysis of burial rates in England from 1770 to 1800. There is a distinct peak in the year of the Laki eruption (Grattan et al., 2005).

extent of the large sulphuric aerosol cloud first described by de Montredon. This cloud reached high up beyond the troposphere into the stratosphere, where very powerful winds distributed its pollutants across the Northern Hemisphere: two weeks after the first eruption, pollutants had indeed already reached almost all of it. This is a first step at attempting to model the climatic consequences of a trap, such as the Deccan. This will be our next endeavour. But, already, we can point out the significant effect on climate and Earth's inhabitants of the modest 1783 eruption of Laki. Compare its 15 cubic kilometres of lava with the volume of some of the largest single flows in the Deccan, which probably exceed 1000 cubic kilometres – that is, almost 100 times larger. And there are hundreds of flows in the Deccan pile. The total original volume of the erupted lava was 200 000 times that of the Laki eruption. The big question is, of course, the exact timing of these flows, and we are working on that. But it seems difficult to escape the conclusion that trap eruptions are a reasonable mechanism for generating massive global climate change and pollution. The evidence is clearer for the Siberian traps (associated with the massive extinction 250 million years ago) than it is for the Deccan. Here lava and ash outcrops at the base of the Siberian traps indicate that the eruptions were either preceded or accompanied by explosive phases during which huge quantities of sulphur were emitted with grave environmental consequences. Sulphur is a key element of cooling, poisoning and acid rain. But carbon dioxide must also be a major player, leading to greenhouse warming. It has recently been realized that fresh lava will quickly be altered, through the formation of carbonates from its constituent minerals which are leached into the world ocean, leading to massive consumption of the excess carbon dioxide emitted by the eruptions (a sort of self-cleansing mechanism, but which may take a million years to operate fully).

Contingency and evidence: has the next catastrophe begun?

There is clearly still a lot of work to do. But I believe the evidence I have tried to summarize here amply supports the idea that flood basalts are the major agent of climate change that have led to the rather (geologically) short and rare events in which evolution of Life on Earth was completely reoriented because of the massive disappearance of large numbers of species. Most of the time, evolution proceeds as Darwin saw it, through survival of the fittest. But there were a few key times when the rules of the game were completely different. It is hard to say that the fittest were those whose life functions were prepared to withstand massive environmental change of a sort that had not occurred on the planet in the previous tens of millions of years. It might be more accurate to speak of survival of the luckiest at these times. Yet, without these few short events, which form the backbone of the geological time scale, Life on Earth would be very different from what it is today. Maybe not in a statistical sense, but certainly species would have been different if the times, locations, durations, chemistry and environment of some ten flood basalts events had been slightly different. Stephen Jay Gould has beautifully summarized the resulting contingency of Life on Earth.

Now for a few concluding remarks. There sometimes is a view among atmospheric scientists who are attempting to predict the results of global change and the future of the environment of Mankind in the coming century that only 'short' time constants are required, and that knowledge of the distant geological past may not be relevant. This is a profound error. On the contrary, increasingly accurate description and understanding of the kinds of catastrophic events that I have described here may be the only way to test the climatologists' models against actual situations which the Earth has already gone through. In more modern words, we may provide the benchmark against which all such models must be tested. And if, as some believe, we have already entered a new episode of mass extinction, possibly due to the very evolution of Man and its impact on the geo- and bio-spheres, it is advisable to understand fully previous episodes which Nature produced by itself, without Man's help. A single large flow in the Deccan traps, erupting over a period of a few years, submitted the Earth to the same kind of stress Mankind has generated in the last two centuries. Geologists may have the best benchmark against which the predictions of climatologists for this century should be tested. Returning to the meanings of 'evidence', I hope the story I have just told you has by now become 'evident', if not obvious, to you. Half of the task of a scientist is to explain the work to others and convince them of the quality and appropriateness of the evidence. But the reader should of course retain a strong sense of critical thinking or considered doubt, remembering that in any circumstance, and particularly the distant past, absence of evidence is not evidence of absence. This problem underlies much of what I have told. I will therefore leave you with a nagging question: where is the next trap going to erupt, and when?

FURTHER READING

Alvarez, W., *T. Rex and the Crater of Doom*, Princeton, NJ: Princeton University Press, 1997.

Courtillot, V., *Evolutionary Catastrophes: The Science of Mass Extinction*, Cambridge: Cambridge University Press, 1999.

Courtillot, V. and Renne, P. , 'On the Ages of Flood Basalts Events', *Comptes Rendus Geoscience* 335, 2003.

Gould, S. J., *Wonderful Life*, New York: Norton, 1989.

Grattan, J., Durand, M. and Taylor, S., 'Illness and Elevated Human Mortality in Europe Coincident with the Laki Fissure Eruption', in C. Oppenheimer, D. M. Pyle and Barclay (eds.), *Volcanic Degassing*, Geological Society of London Special Publication 203, 2003.

Grattan, J., *et al.*, 'Volcanic Air Pollution and Mortality in France 1783–1784', *Comptes Rendus Geoscience* 337, 2005.

Keller, G., 'Chicxulub Impact Predates the K–T Boundary Mass Extinction', *Proceedings of the National Academy of Sciences* 101, 2004.

Thordarson, T. and Self, S. , 'Atmospheric and Environmental Effects of the 1783–1784 Laki Eruption: A Review and Reassessment', *Journal of Geophysical Research* 108, 2003.

Ward, P., *The End of Evolution*, New York: Bantam Doubleday Dell, 1994.

4 Evidence for Life beyond Earth?

MONICA M. GRADY

Introduction

The question of whether or not there is life beyond Earth is one that has been asked many times throughout history. The possibilities of extraterrestrial life have, until recently, mainly been addressed by film-makers and writers of science fiction – pictures of multi-limbed (inevitably hostile) aliens invading Earth occur frequently. In the last decade, however, discussion of extraterrestrial life has moved from science fiction to mainstream science. The reason that there is now an apparent acceptance of the likelihood of extraterrestrial life is because of the images and data that have been returned by a variety of scientific instruments, both ground- and space-based. Telescopes have discovered over 100 planets orbiting other stars, opening up the possibility that, perhaps in the not too distant future, we will observe planets that have similar characteristics (atmosphere, surface water, etc.) to Earth. Within the Solar System, spacecraft orbiting planets and their satellites have sent back detailed pictures of these other worlds. On our neighbouring planet Mars, we have seen volcanoes, craters, cliffs and networks of valleys. Further away from Earth, instruments on the Galileo satellite that orbited Jupiter recorded signals that suggest there is a deep ocean below the icy crust of Jupiter's satellite Europa. Back on Earth, there has been increased awareness of the range of habitats that are occupied by micro-organisms. The physical limits (temperature, acidity, etc.) at which life can exist extend from sub-zero to temperatures well above that of boiling water; microbes can also survive in environments of extreme acidity and alkalinity. Analysis of one particular meteorite from Mars found preservation of a possible fossilized bacterium deep inside the meteorite, suggesting that survival of micro-organisms might even have extended to Mars. It

73

now seems that micro-organisms can inhabit environments previously thought to be too inhospitable to support life.

Perhaps because of the inspirational nature of satellite, telescope and microscope images, or, more mundanely, because shortfall in funding requires scientists to forge new alliances, there is a new population of scientists willing to cross traditional subject boundaries, using materials and evidence from related fields to enhance knowledge and understanding of their own interests. Thus has the science of astrobiology been revived. Astrobiology may be summarized as the study of the origin, evolution, adaptation and distribution of past and present life in the Universe. The search for life in the Universe has to start with an understanding of life on Earth – how, where and when did life emerge on Earth, and where might these conditions be duplicated elsewhere in the Solar System, and beyond. In this chapter, I will firstly review the evidence that life *could* exist beyond Earth, before assessing the evidence that might indicate that life does (or did) occur outside our planet.

How is Life defined?

Before launching into a search for life beyond Earth, we must first be able to define what life is. It is difficult to find a single completely satisfactory definition of life: different groups of scientists describe what they mean by life in different ways. At school, students are taught that there are seven characteristics of living beings: growth, reproduction, respiration, nutrition, excretion, locomotion and response to external stimulus. However, this is by no means a complete description of something that is alive. For instance, the properties of fire might all be fitted into this description: as long as there is fuel (nutrition), a fire will grow, move, take in air, give out heat, respond to wind direction and so on. But we would never describe a fire as being 'alive'. A chemist might define a living organism as a lipid-encapsulated system comprising mutually interdependent (homochiral) nucleic acids and coded peptides/proteins. But this description might also fit a virus. A virus is a microscopic organism that consists of a core of nucleic acids surrounded by a sheath of proteins. But would we describe a virus as alive? Although a virus has the potential to grow and reproduce, it does not contain the necessary cellular components to do so by itself: it needs a host. Viruses are parasites – they are living organisms, but cannot sustain life independently from their hosts. In any search for life, it is unlikely that we would find viruses alone – they would always be associated with their host organism.

Defining life in this sort of way is a fairly standard scientific approach. There are, however, other constituencies who have very different criteria by which life is defined. For example, early philosophers, including Aristotle, attempted, on the basis of logical deduction, to describe the distinctive and irreducible characteristics of living things. The theological viewpoint requires a deity in order to give meaning to life. The thirteenth-century theologian Thomas Aquinas wrote that 'a living thing is more perfect than what merely exists, because living things also exist and intelligent things both exist and live'. Aquinas here draws a distinction between 'existence', in which a being has no purpose, and 'living', in which a being has the purpose of reaching perfection, perfection being God. As is the case today, early arguments were based on the available evidence, and discussed in the framework of the understanding of the time. Looking more to the future than the past, progress in information technology is already starting to blur the distinction between natural and artificial intelligences. The development of advanced neural networks, information processing systems modelled on the way that biological systems process information, has already resulted in computer programs that can learn, adapt and evolve. Can we envisage a time, perhaps in the not too distant future, when it is difficult to distinguish between biological and abiological life? And if that is the case on Earth, how will we make a similar distinction if we discover advanced life forms beyond the Solar System? Is it an important distinction to make? So, although advances in knowledge resulting from technological developments have given us more evidence on which to base our views about the origin of life, we are still striving to come to the ultimate definition of 'life'.

A full discussion of the theological, metaphysical and artificial intelligence aspects of the definition of life is beyond the scope of this essay. I have included the above brief comments as an indication that what might seem apparent to one group of investigators is by no means self-evident to another group, observing the same problem from a different perspective. But because I am treating the origin of life from the viewpoint of a planetary scientist, I will focus my discussion on generally accepted scientific hypotheses, and not stray into regions where more metaphysical arguments might be employed. Returning, then, to a biological designation for life (and assuming that, at least within the Solar System, we are not likely to meet any advanced neural networks masquerading as biological entities), and drawing inferences from the living/non-living examples of viruses and fire, I will take the ability to adapt and to evolve independently as being necessary characteristics of viable living beings. As an organism, or group of organisms, evolves, information about the organism must be transferred from one generation to the next. This transfer of information is 'heredity', and the process of inheritance is another way of recognizing something that is alive and capable of adapting and evolving.

What sort of evidence should we look for?

Now that we have made an attempt at defining life as an information system capable of inheritance, adaptation and evolution, we must give some thought to what living organisms look like, and what we should be searching for. Currently, there is vigorous debate amongst taxonomists about how living organisms are best described – by their external morphology (e.g., it has got leaves and yellow flowers) or by its genetic makeup (a long sequence of ACTG, etc.). Carl Sagan encapsulated part of this dilemma when he wrote 'the search for extraterrestrial life must begin with the question of what we mean by life. "I'll know it when I see it" is an insufficient answer.' On Earth, life comes in a variety of forms: with legs, wings, fins, fur, flowers and roots. These are all examples of complex life forms, and are the results of evolution over the past 640 million years or so. They leave abundant evidence for their existence in rocks: fossils such as ammonites and dinosaur bones are readily recognizable as a record of once-living organisms. But we know already, from orbiting spacecraft, that there are no trees, dinosaurs or higher mammals on any of our neighbouring planets, so if life is, or has been, present, it must occur, or have occurred, as much less complex organisms. This would accord with the spread of life on Earth, where the most numerous and widely distributed of living entities are not eukaryotes (macrofauna or flora made up of complex cells with a nucleus, etc.). The biomass of our planet is dominated by bacteria and archaea. These are prokaryotes, less complex organisms with simple cells devoid of organized elements such as a nucleus, and they leave very little trace of their presence in the fossil record.

In order to uncover life's origins, we must go back to the origin of Earth. We have a good idea of how Earth formed, from a rotating disc of gas and dust ~4567 million years ago, as part of Solar System formation. But although the picture of this process is very detailed, and based on inferences drawn from study of asteroids, comets and meteorites, the picture remains a hypothesis that may never be absolutely proven, simply because we were not there to witness the formation directly. However, as our ability to detect and observe newly forming stars and planets becomes ever more proficient, then perhaps we will soon be able to see whether our hypothesis of planet formation through dust agglomeration and coagulation is correct.

Notwithstanding uncertainties in details of Earth's formation, it is likely that, once Earth formed, its surface was hostile to life for at least the first 350 million years or so, as the Solar System settled into the structured geography that it has today. Again, we have no absolute evidence for the processes occurring during this earliest epoch of Earth's history when the planet was bombarded by comets and asteroids, keeping the surface molten. No trace of the original surface remains – we can only infer the extent of bombardment by observing the number and relative ages of craters on our close neighbour, Moon, which show that our satellite was subject to a heavy bombardment. It is not known whether, between episodes of bombardment, conditions were such that life might have arisen and then been wiped out. This might have happened several times as the Solar System gradually achieved a more stable state, and Earth's surface cooled and solidified. What we do know, from the occurrence of specific mineral species, is that by only $\sim$350 million years after its aggregation, Earth's surface was sufficiently cool to allow water to condense, and for there to have been rivers and seas. Because of Earth's turbulent early history, until relatively recently it was not known whether life emerged once, in one place, from which all other organisms were derived, or whether life forms evolved more than once, with elimination or extinction of life occurring during periods of enhanced impact frequency. Advances in isolating small fragments of DNA and RNA have allowed determination of the genetic make-up of organisms, leading to construction of a 'Tree of Life', in which all organisms stem from a common ancestor. This implies a single origin for life, rather than several independent origins.

As Earth gradually cooled, oceans formed. But Earth was still very different from today: its atmosphere was reducing in nature, comprising hydrogen (H_2), methane (CH_4) and carbon monoxide (CO). The earliest organisms were anaerobic photosynthetic organisms such as cyanobacteria. Gradually, the balance of gases in the atmosphere and the deep ocean changed: oxygen built up with the photosynthesizing activity of the bacteria, and carbon dioxide abundance decreased, owing to the formation of limestones and the weathering of silicate rocks. By around 2 billion years ago, the atmosphere had switched from reducing to oxidizing, with oxygen (O_2) dominating over carbon dioxide (CO_2). By the start of the Cambrian period, 543 million years ago, multi-cellular organisms were abundant, implying a flourishing biosphere, and macrofossils started to appear regularly in the fossil record.

We still do not know whether life on Earth first arose in surface waters, or at depth. Searching for evidence for ancient Life on Earth is difficult – the fossil record is incomplete, and the very earliest micro-organisms would not necessarily have left fossils. The oldest traces of biological activity on Earth were thought to have been preserved in rocks from the 3.85-billion-year-old Isua Complex in West Greenland, indicating that life could have arisen on Earth almost as soon as it was possible for it to do so. The tracers are 'chemical fossils', so-called because they were identified by changes in the chemistry of the rocks that can only be detected using sensitive analytical techniques. However, interpretation of the 'chemical fossils' is not straightforward, and there is not unanimous agreement that the observed changes in chemistry were caused by biological processes. A similar debate surrounds the oldest physical fossils, which for many years were claimed to be 3.5-billion-year-old remnants in the Apex Chert of the Warawoona Formation in Western Australia. Recently, it has been suggested that they are not fossils, but patterns left in the rocks by fluid activity. Stromatolites are laminated algal mats; features present in 3.2-billion-year-old Archaean rocks in Western Australia have been interpreted as fossilized stromatolites, but again have recently been re-interpreted as chemical precipitates. As analytical and observational techniques become more refined and sophisticated, newly recognized features and characteristics refute what had been taken as firm evidence of ancient life. It now appears that the oldest undisputed physical fossils are only about 3.1 billion years old. I mention this here to illustrate the difficulty of identifying ancient fossils. If we cannot do so with certainty on Earth, where the locality at which the material was found can be revisited, and the features examined and measured by every conceivable analytical technique, then what chance is there of detecting evidence for fossilized life on other planetary bodies? Given that ancient trace fossils are not abundant in terrestrial rocks and, as discussed above, are very difficult to identify with confidence, it is perhaps a bit of a long shot to hope that a space probe would be able to detect fossilized ancient non-terrestrial life. Is there anything else that we could use to guide our searches? One possible solution is to search for trace gases in a planet's atmosphere, and I will return to this in a later section.

Conditions for life

To explore evidence for life beyond Earth, we must set ground rules, to enable the search to retain some connection with the physical realities of our current environment. If we do not set such rules, then we are in danger of allowing the search to become a flight of fancy, soaring away into the realms of science fiction. Our starting assumption, then, is a very basic one: that the laws of physics and chemistry are not violated on bodies beyond Earth. Certainly, within the Solar System, and probably within the entire measurable Universe, there is no reason to suspect that these laws would not be obeyed. (Caveat: I state *measurable* Universe; in localities such as black holes, I am not sure what sort of physical or chemical laws, if any, would hold.) Thus the usual atomic structure (a positively charged nucleus of protons and neutrons, surrounded by a cloud of electrons) is the framework within which chemical processes are constrained. Following on from the assumption of the universal applicability of physical and chemical laws is our second assumption – that any life form we find will be carbon-based. This is perhaps a more provocative assumption, but I will attempt to justify it.

The carbon (C) atom is unique in its ability to form a vast range of compounds. This property is a result of carbon's atomic structure, which allows it to form very stable covalent bonds. Carbon is the only atom that can form long chains by bonds to other carbon atoms (a class of compounds known as aliphatics). Carbon also forms rings of six atoms (aromatic compounds). The rings can join together; side chains of aliphatic compounds might also be attached, leading to large and complex molecules. The chemistry of carbon compounds is organic chemistry. (Note: not biogenic, which implies life-bearing, but organic, meaning carbon-bearing.) The only other atom that can approach carbon in its variety of compounds is silicon (Si), an element within the same group of the periodic table as carbon, and so related in its atomic structure. However, silicon cannot form extended chains or rings by itself – it can only produce structural arrays when combined with oxygen (O). In the same way that the C–C bond is the foundation of organic chemistry, the Si–O bond is the basis of rock-forming minerals on Earth.

One of the dogmas surrounding discussions of the origins of life is that water is essential for life. This is rather misleading: the presence of water is not a necessity – but the presence of a suitable solvent is. At the cellular level, a solvent is required for several purposes: to dissolve and transport salts (nutrients); to enable removal of waste products; to provide rigidity to cells and cell walls. The properties of water (a polar solvent, stable over the same temperature range as pertains at Earth's surface) are such that it is the most versatile solvent we have on Earth. And so liquid water is deemed to be necessary for life on Earth and, by extension, necessary for carbon-based life elsewhere.

Stages in the origin of Life

So then, we are assuming that physics and chemistry are universally constant, and that life is carbon-based and requires water. These assumptions allow us to make comparisons between what we know (Life on Earth) and what we are looking for (Life beyond Earth). The assumptions provide a baseline against which observations can be judged; without such a reference point, we would be in danger of hypothesizing beyond rational limits. We need now to make yet another assumption: that the *process* that produced Life on Earth operated throughout the Solar System. We infer that the process that produced life occurred in stages, starting with chemicals, and building up to organisms. What were the chemicals from which life arose, where did they come from and what else was necessary before life could form?

The early Solar System was a dynamic and unsettled environment. As stated above, the early Earth was a hostile place: its surface was bombarded frequently by asteroids and comets, causing it to remain molten for many millions of years after aggregation. During this epoch, Earth out-gassed, or expelled, a significant fraction of the volatile elements (hydrogen, carbon, nitrogen, etc.) that had been a part of the primordial dust from which it formed. Gradually, bombardment within the inner Solar System died down, and the Earth's surface cooled and solidified. Impacts lessened in frequency but did not cease entirely and in this period the impactors added volatiles to Earth, rather than contributed to their removal. The main agents of volatile delivery were the dusty iceballs that are comets. Given that they traverse the Solar System, it is likely that they also delivered volatiles to other planets. Note that here I am talking about the transport and delivery of inorganic species throughout the Solar System. I am

not considering the movement of living organisms through space. That theory –
the theory of 'panspermia' – was constructed formally by the chemist Svante
Arrhenius in the early twentieth century. Arrhenius proposed that microscopic
spores were blown through interstellar space by stellar radiation pressure, eventually seeding Earth with life. In recent years, this theory has been tested; it
is now known that some species of bacteria can survive enormous radiation
fluxes, especially when carried within small fragments of rock. However, even
if panspermia were shown to be feasible across interstellar as well as interplanetary distances, it does not explain the *origin* of life – it merely places the site of
life's source in a different region of space: whence did the interstellar bacterial
travellers set out? I will return to the more local interplanetary transport of
bacteria in a later section.

It is generally believed that Life on Earth started from simple molecules
(carbon monoxide, methane, hydrogen sulphide, ammonia). Simple chemical
species reacted together to produce more complex molecules. How, and where,
did this occur? This simply is not known. There have been suggestions that
the reactions occurred in the atmosphere, with lightning bolts as the energy
source. Charles Darwin himself had his own ideas as to how and where life
might have started: 'But if (and oh! what a big if!) we could conceive in some
warm little pond, with all sorts of ammonia and phosphoric salts, light, heat,
electricity, etc. present, that a protein compound was chemically formed ready
to undergo still more complex changes' (Charles Darwin, letter to J. D. Hooker,
1871). Simulations undertaken in the early 1950s, in which amino acids were
produced by discharging a spark through a mixture of gases, provide some
evidence that this is a reasonable surmise, but the mechanisms required to
build more complex molecules into extended chain molecules such as RNA and
DNA are still poorly understood. Analysis of large organic molecules shows
that many have a hydrophilic and a hydrophobic end. Assemblages of such
molecules are likely to lead to their alignment, in turn allowing them to form
surfaces that either repel or attract water molecules. In this way, perhaps,
surfaces could become membranes, separating molecule assemblages from the
solvent in which they formed. The formation of a template for self-replication
is the final step that brings chemistry to its interface with biology, and the start
of evolution.

There are philosophical issues that accompany this very sweeping and non-
specific outline of the processes leading up to the formation of life, most

particularly the distinction between 'alive' and 'living'. I have already put forward as a definition of life a system that can adapt and pass on inheritance; we might now run into debates about whether or not certain organisms are (merely) alive or in fact conscious (living). At what stage in evolutionary history did organisms develop consciousness? These discussions hark back to arguments that followed Darwin's publication of *On the Origin of Species* in 1825, arguments over such questions as whether or not dogs possessed souls. Such issues are fundamental to our understanding of life and its development on Earth, and are by no means resolved. Given that, how much more difficult will the recognition of conscious life forms be on planets beyond Earth? We have not yet reached a consensus as to how we would recognize something that has life, let alone been able to mark a division between consciousness and lack of it.

The biological envelope on Earth

The limits within which life can survive, grow and evolve are its 'biological envelope'. These limits are based on the physical properties of the compounds that make up organisms. Because we have made the basic assumptions that the laws of physics and chemistry hold when we seek life beyond Earth, and because we have no other frame of reference within which to work, we use the parameters of the biological envelope to guide the search for life elsewhere in the Solar System. The most obvious signs of Life on Earth are the macrofauna and -flora – all examples of complex species, and almost all of which occupy environments of moderate temperature, etc. As outlined above, the greatest proportion of Earth's biomass is made up of micro-organisms, from the archaea and bacteria kingdoms. These less complex organisms occupy a vast range of environments, and it is their ability to survive extreme conditions of temperature, etc. that defines the biological envelope of Life on Earth. In the time between the emergence of life and the present day, microbes have colonized every niche on Earth in which it is possible for life to survive. Those that exist in the most extreme physical environments are known as 'extremophiles'. Currently, the highest known temperature at which an organism can survive is +121 °C, and the lowest temperature at which a metabolism has been detected is –17 °C. Acidity levels from pH 0 to 12.5, salinity up to 5.2 M and pressures in excess of 100 MPa are environments (or biotopes) in which microbial life has been found to survive. (For comparison, the mean surface temperature of the North Sea in summer is 15 °C, its pH around 7, salinity 0.6 M and its mean atmospheric pressure 0.1 MPa.) When under extreme conditions, micro-organisms have had to evolve protection strategies to enable their survival. The ability of organisms to survive in high-temperature biotopes is limited partly by the stability of liquid water, and partly by the molecular stability of constituent organic molecules. If the temperature at which organic compounds break down is exceeded, then the organism will not survive unless the components have been stabilized by additional ions or functional groups. At the other end of the temperature spectrum, psychrophiles can survive down to −17 °C. There are eukaryotes that can survive down to these low temperatures, as well as archaea and bacteria; such species include fungi. Survival strategies developed to mitigate the effects of low temperature include the development of lipid-based systems. Psychrophiles have shown the ability to survive for thousands of years: ice cores from Lake Vostok (see below), where the ice is up to 200 000 years old, contain unusual microbes that are viable when cultivated.

Habitable niches on Earth

Direct observation of the different biotopes in which life exists on Earth acts as a guide to where life might survive in niches beyond Earth. So we can consider environments colonized by micro-organisms on Earth, and predict where else life might have arisen in the Solar System, given similar environmental conditions. Observations by spacecraft of planets and their satellites have shown that there are no known planetary surfaces on which liquid water is stable, or where moderate temperatures of around 20 °C pertain. Conditions beyond Earth are extreme: temperatures range from >400 °C at the surface of Venus to <−200 °C on Saturn's giant satellite, Titan. There are similar extremes in terms of atmospheric pressure and radiation flux. So we assume that any life present beyond Earth will have to have adaptations to extreme environments as do the terrestrial microbes that survive in extreme environments on Earth. On this basis, I will discuss two extreme environments on Earth that have been studied in the context of their relevance to conditions that might exist elsewhere in the Solar System.

At one extreme lies Antarctica. Its cold desert climate cannot support a complex ecosystem of plants and animals. Even so, there is a significant biomass of psychrophiles within the sandstones and quartzites of the Dry Valley regions. Communities of lichens, known as cryptoendoliths (literally, 'hidden inside the rock'), colonize the top layer of the sandstones. Despite the dry conditions of the Antarctic environment, the microscopic communities thrive on water trapped within pore spaces in the sandstones. External temperatures are frequently as low as $-30\,°C$, but the warmth generated within the sandstones can reach $10\,°C$. Dark pigments in the microbes block out excessive sunlight and absorb UV radiation. These communities have successfully adapted to survive in a cold, windy environment, with reduced water availability. Depletions of the atmospheric ozone layer, brought about through terrestrial pollution, have resulted in a hole in the ozone layer centred over the South Pole, leading to higher-than-expected solar UV radiation fluxes. Comparison of this environment with that at the surface of Mars shows many similarities, although, of course, Mars has an even thinner atmosphere than that of Earth. Even so, the rock-dwelling bacteria are considered to be useful models for the types of biota that might be found within rocks at the Martian surface, a cold, dry and windy environment similar to that of the Dry Valleys. The big challenge, one that none of the spacecraft that have landed on Mars has been able to address, is how to test for or recognize such a biota were one to exist at Mars' surface.

A second Antarctic ecosystem is being studied for the relevance that it might have to potential life forms on Jupiter's ice-covered satellite Europa. Lake Vostok is the largest of the 80 sub-glacial lakes in the Antarctic plateau, and is covered with layers of ice up to 4 km thick. The deepest ice-layers are up to 400 000 years old, and have presumably been isolated from the outside environment since being laid down. Drilling through the ice (but not into the lake) has produced cores in which viable bacteria, fungi and algae have been discovered. The species were sealed in the ice, and are different from related species extant at the surface today. Because of the sub-glacial nature of Lake Vostok, it is believed to be a useful terrestrial analogue for Europa, indicating that the survival of organisms in such environments may be possible.

A very different extreme of environment exists at places on the ocean floor. Molten rock wells up and forms new oceanic crust at spreading centres along ocean floor ridges. Hydrothermal vents were discovered close to these areas in the late 1970s. The vents, or 'black smokers', are hot springs where super-heated water (up to $350–400\,°C$), rich in hydrogen, methane and hydrogen sulphide, shoots up from the sea floor. Where the hot water meets the cold, oxygen-rich bottom water, there is an instant chemical reaction and sulphides precipitate out from the water, colouring it black. The sulphides build up rapidly to 'chimneys' reaching heights of several tens of metres. Discovery of the vents revealed that, despite the depth and darkness, parts of the ocean floor are home to an unusual collection of animals such as mussels, crabs and tubeworms, feeding on the hyperthermophilic bacteria and archaea that flourish in these very hot conditions. The discovery of a successful ecosystem based on chemical energy rather than photosynthesis has raised the possibility that life may not have arisen in surface waters, as original theories purport. Discovering communities entirely supported by chemosynthesizing microbes has given impetus to the search for life in other deep oceans, especially on Jupiter's satellite Europa, where a liquid water ocean occurs below the visible crust of ice.

Habitable niches beyond Earth

During the formation of the Solar System, Earth suffered intense bombardment by asteroids and comets, the latter in particular delivering the water and organic compounds considered to be the crucial ingredients for life. If comets and asteroids could deliver such materials to Earth, then they could also deliver them to other planets and their satellites within the Solar System. Therefore there is the potential for life to have started in other habitable niches throughout the Solar System. Knowledge of where micro-organisms flourish, and the wide range of habitats occupied on Earth, helps to pinpoint where life might exist elsewhere within the Solar System. The two most likely candidates identified so far are Mars and Europa.

Evidence for life on Mars?

Ever since 1877, when Giovanni Schiaparelli described *canali* on the surface of Mars, our neighbouring planet has been regarded as a probable home for alien life. The early-twentieth-century astronomer Percival Lowell spent a major part of his life looking for evidence for life on Mars. He was convinced that the *canali* were artificial channels built by Martians. He wrote 'that Mars is inhabited by beings of some sort or other we may consider as certain as it is uncertain what those beings may be'. However, although it is now understood that the markings Lowell observed were clouds or dust storms, and not surface features, pictures taken more recently by orbiting satellites have shown that there are indeed channels (not artificial canals) on Mars' surface. Images returned by the Mariner 9 orbiter mission of 1971–2 revealed channel and valley networks that bore striking similarities to river and stream features on Earth. These were assumed to demonstrate that Mars must have had water running across its surface at some time in its history. The most recent high-resolution images of Mars' surface come from ESA's Mars Express orbiter, and show the channels and valley networks in even greater detail. Some surface features have been re-interpreted as it is believed that either the channels might have been formed by ice, rather than liquid water, or the flow was below the surface of the planet. Whatever form it took, the implication is still that water was relatively abundant, perhaps as recently as 1 million years ago. NASA's Pathfinder mission of 1997, and its Spirit rover of early 2004, recorded spectacular images of a rock-strewn plain, showing rounded pebbles and layered structures also consistent with the presence of water at times in the past. The landscape of part-rounded pebbles and boulders could also be interpreted as evidence of catastrophic flooding. In contrast, the Opportunity rover (the twin of Spirit, which also landed in early 2004, but in a different region of Mars) showed pictures of a nearly featureless plain. But close-up analysis of a low outcrop showed that evaporates occurred, and that the surface soil was rich in tiny spherulites, characteristic of fluid flow. All these observations indicate that water was once fairly abundant and widespread across Mars' surface in the past. Data from the neutral mass spectrometer aboard the Odyssey spacecraft, which arrived in orbit around Mars in 2001, indicate that water or ice still occurs in sub-surface locations today. For water to be present on the surface of Mars in the past, the atmosphere must have been much thicker and surface temperatures much warmer than they are today. A thicker atmosphere ensures greater protection from solar radiation, and with wetter conditions, all the requirements necessary for the emergence of life, in theory, would have been in place.

There are other indicators that Mars could have been habitable in the past. Direct analysis of Martian meteorites has revealed different types of salts produced by aqueous processes. Halite has been found, suggesting that brine pools must have been relatively common on the surface of Mars. Rapid desiccation of the surface after loss of its atmosphere may have rendered Mars rich in regions of high salt concentration, in which halophilic micro-organisms might have flourished. Carbonates also occur in Martian meteorites, and almost certainly formed on Mars as the result of the slow evaporation of warm water locked in enclosed basins. This is an environment in which, as has been seen on Earth, micro-organisms may be sustained. Despite the surface water having now dried up, it is possible that liquid still resides in the sub-surface soil layers, housing bacteria. Alternatively, micro-organisms might inhabit the rocks themselves, as in the Antarctic cryptoendolith communities. Shapes found embedded in carbonate patches in a meteorite in the Allan Hills (ALH) 84001 Martian meteorite led NASA scientists to claim in 1996 that they had discovered evidence for Martian microfossils: 'we conclude that they are evidence for primitive life on early Mars' (D. S. McKay *et al.*, *Science* 273, 1996, 924). The claim, however, remains highly controversial and is still the subject of much debate.

Recently, there has been much discussion on the exchange of material between planets. We have meteorites from Mars, so we know that rocks can travel from there to Earth. It is also possible, but less probable given the less favourable orbital and gravitational dynamics, that material can travel from Earth to Mars. We also now know that some bacteria can survive high radiation doses and low temperatures. These observations have led to the speculation that Mars could have been seeded by life from Earth, or vice versa. Unfortunately, we have no evidence for this, no matter which way the microbes were posited to have been travelling, and we will not be able to determine the likelihood (or otherwise) of interplanetary panspermia unless and until we uncover life on Mars. Assuming we do find life on Mars, and that it is organic and DNA-based, there remain four possibilities: (1) life arose on Earth and seeded Mars; (2) life arose on Mars and seeded Earth; (3) life arose on Mars and Earth independently; or (4) life came from elsewhere and seeded both Earth and Mars. At the moment, we do not have evidence to distinguish between these origins. The question of life on Mars is one of the major driving forces behind the current Martian exploration programmes of all the main space agencies.

Evidence for life on the satellites of Jupiter and Saturn?

Europa, one of the four Galilean satellites, is similar in size to Earth's Moon and is judged to possess a range of properties and environments that maintain the potential to harbour life. Information and images from NASA's Galileo mission (orbiting Jupiter and the Galilean satellites between 1995 and 2003) indicated that Europa is composed of metal and silicates, but has an icy crust. Sunlight reflecting from the ice makes Europa's surface one of the brightest in the Solar System. Magnetic data have been interpreted as evidence of a sub-surface salty ocean beneath a thinner water–ice shell. The water is kept liquid by internal heat generated from the combined gravitational influence of Jupiter and the innermost Galilean satellite, Io. Europa, therefore, possesses the fundamentals necessary for life: water and energy. There has been speculation that Europa's ocean might be heated from the bottom upwards by hydrothermal vents similar to those found on Earth's ocean floors. If this is the case, then, as on the Earth, these vents might host a rich variety of organisms.

Titan is far and away Saturn's largest satellite, and is the second largest satellite in the Solar System (after Jupiter's Ganymede). Titan is enveloped by a thick atmosphere, the main component of which, as on the Earth, is nitrogen (Figure 4.1). However, unlike Earth, the remaining constituents are methane and argon, with trace amounts of higher hydrocarbons (ethane, propane, ethyne, etc.), carbon monoxide, hydrogen cyanide and cyanogen. Radar measurements from Earth, coupled with observations from the Hubble Space Telescope and recent images from the Cassini spacecraft, have built up a picture of Titan as having a rough rocky and icy surface with a temperature $\sim -200\,°C$. Condensation of ethane from the atmosphere has resulted in the build-up of lakes and ponds of ethane in which ammonia and methane are dissolved. Although there is no suggestion that the environment of Titan is in anyway suitable to harbour life, the atmospheric composition and surface conditions are sufficiently primitive for them to have been taken as possible analogues to the conditions extant on the early Earth, prior to development of the biosphere. Hence, understanding the atmospheric and surface processes on Titan is a key to understanding the processes that led to the evolution of Life on Earth. As a consequence of this significance, Titan was the target of Huygens, an ESA-led probe, that landed on the surface of the satellite in January 2005 (Figure 4.2). The probe descended through Titan's atmosphere for several hours, recording its elemental and isotopic composition, wind speed and temperature. The probe took several images as it descended through the clouds, and, once it had landed, took a single image of a flat plain strewn with rounded boulders of ice. It is not yet certain whether some of the features seen in the images are seas of ethane, or rocky plains.

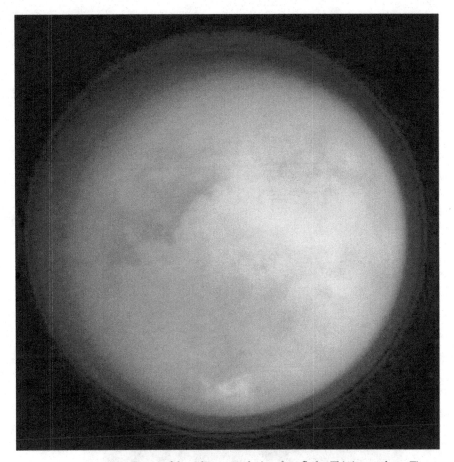

FIGURE 4.1. Titan in false colour, seen during close fly-by. This image shows Titan in ultraviolet and infrared wavelengths. It was taken by the Cassini–Huygens Imaging Science Subsystem on 26 October 2004, and is constructed from four images acquired through different colour filters. Credit: NASA / JPL / Space Science Institute.

Evidence for life beyond the Solar System?

Several space missions are planned before 2017 that may shed more light on the presence, or otherwise, in the Solar System, of life beyond Earth. Up to the end of 2004, results from telescope surveys had identified around 125 extra-solar planets, giving rise to increased speculation about the possibilities of life beyond the Solar System. Unfortunately, only a few types of stars are likely to

FIGURE 4.2. Image of Titan's surface taken by the Huygens probe as it landed on the surface. Sizes have been added to indicate the scale of the features. The surface is darker than originally expected, consisting of a mixture of water and hydrocarbon ice. There is also evidence of erosion at the base of these objects, indicating possible fluvial activity. Credit: European Space Agency (ESA) / NASA / JPL / University of Arizona.

support planetary systems, and thus be potential hosts for life. There are several parameters that are important in determining whether or not a star is likely to be at the centre of a planetary system, including the age, size and stability of the star. So although there are many billions of stars in our Galaxy, many of them are types (too old, too big, too small, too energetic) that are unlikely to support planetary systems for sufficient lengths of time for life forms to arise or evolve. But we should not be too pessimistic about the chances of finding life beyond the Solar System. As Freeman Dyson wrote in 1979: 'I conclude from the existence of these accidents of physics and astronomy that the Universe is an unexpectedly hospitable place for living creatures to make their home in.' There are still plenty of places in which we can search for evidence of life.

On the basis of the parameters detailed above, the search for life outside our Solar System should be directed towards Earth-like planets, orbiting Sun-type stars at distances considered to be within the Habitable Zone of the observed star, which itself should be located within the Habitable Zone of its galaxy. The techniques and instrumentation needed to observe planets outside our Solar System have been developed relatively recently. The first observation of a planet orbiting a Sun-type star was made in 1995, when a large body was detected orbiting the star 51 Pegasi. Since then, about 125 giant planets have been detected around nearby Sun-type stars, and more are detected each month. So far, none of the planets discovered has had the characteristics that would designate it as a likely candidate for further study as a host for life: they have all been Jupiter-like gas giants orbiting close to their stars. The next stage in the search for extra-solar planets can only come with technical improvements. Both the ESA and NASA have missions in the planning stage (Darwin, Kepler and Terrestrial Planet Finder) which will have the capability both to detect Earth-like planets and to determine their characteristics.

Signatures of life

The signature of Life on Earth is indisputable – any spacecraft passing in close proximity to our planet would detect evidence for life (abundant water and organic material), and even of intelligence (regular thermal and radio emissions). Unfortunately, such clear indication of life on other Solar System bodies is very unlikely. What should any mission to detect extraterrestrial life be looking for? Signatures might remain as traces in an atmosphere, in surface layers or in sub-surface regions; all possible niches must be examined. For Life on Earth to succeed, the importance of water and organic compounds has been stressed; any search for extraterrestrial life is therefore likely to include, at the very least, the search for water and carbon.

The relevant signatures fall into two groups: those that can be measured directly, and those that are observed remotely. In the former category, measurements of organic compounds are likely to be of the greatest significance. The simple presence of organic molecules is, of course, insufficient evidence to infer a biota, since organic compounds can have a purely abiogenic origin: they are found in comets and meteorites, on Titan, etc. There are, however, two analytical techniques that can be used to distinguish the by-products of biological processes if organic molecules are detected *in situ*. The first method measures the optical chirality of the molecules. On Earth, stereoisomers of an organic compound (e.g., the amino acid alanine) produced abiologically generally form a racemic mixture (i.e., contain an equal mixture of the D- and L-enantiomers), whereas the same compound produced biologically is present as only the L-enantiomer. It is possible that a similar effect might occur with extraterrestrial biology, although there are some astrophysical environments in which enantiomeric excesses can be generated, e.g. irradiation of presolar material by circularly polarized ultraviolet light.

The second method for determining the biogenic, or otherwise, nature of extraterrestrial organic compounds uses the isotopic composition of the material as a marker. Non-biological systems tend towards equilibrium. Biology introduces disequilibrium, for example during carbon fixation the isotopic composition of carbon becomes fractionated, such that the ratio of ^{13}C to ^{12}C is lower in the product relative to the starting material. As a result, biologically produced organic matter is depleted in ^{13}C relative to the source carbon. Assuming an optimistic scenario, when searching for life, say on Mars, the final products might be waste products, or fossilized remains that have different carbon isotopic signatures from the inorganic carbon also present in the soil, and from carbon dioxide in Mars' atmosphere. The GAP experiment, designed to search for just such a disequilibrium isotopic signature in soils excavated from below the Martian surface, was part of the Beagle 2 lander that, sadly, was lost on its descent to the Martian surface in December 2003.

Fortunately, since it is not always possible to analyse planetary materials directly, there are indicators of life for which searches can be made using remote observation techniques. The most common tool is the spectral signature of a planetary body, most usually of its atmosphere. If there were no life on Earth, then the planet could be described as a system with three interacting (and inter-dependent) reservoirs: the atmosphere, lithosphere and hydrosphere. As the planet aged, the three reservoirs would come to internal quasi-equilibrium, disturbed episodically by external factors such as asteroid impacts, etc. Gases in the atmosphere would also be in equilibrium, with species such as methane (out-gassed from volcanoes) removed by photolysis almost as soon as they had been produced. Species such as oxygen are also unstable against UV radiation, and are broken down on short timescales. Add a fourth reservoir to this system, a biosphere, and immediately the system is perturbed and does not achieve equilibrium. Some organisms produce methane, others oxygen, and the concentration of these gases in the atmosphere are constantly replenished by biological activity. A satellite observing a living planet would see an atmosphere out of equilibrium, containing a combination of gases (methane plus oxygen/ozone) that would not exist on a planet with no life. So one of the objectives of planet hunters is to search for planets with atmospheres in disequilibrium.

In the search for atmospheric disequilibrium, the presence of ozone (which implies oxygen) or methane would be most diagnostic – both molecules are rapidly removed from the atmosphere, and so their signature would indicate a source of continual renewal from a biological source. Future searches for Earth-like planets (ELPs) will concentrate on the detection of conditions that would indicate environments conducive to the survival of life. Thus infrared spectrometers will be employed to measure the thermal emission characteristics of ELPs, to determine surface temperatures, and the presence or otherwise of water. The utility of these techniques was shown when the *Galileo* spacecraft, en route to Jupiter, flew by Earth and found evidence for life. There is a huge problem, though, with this line of reasoning. It is assuming that any biosphere is aerobic, i.e. requires oxygen to survive. For the first 2.2 billion years of Earth history, our atmosphere contained no free oxygen, and our biosphere consisted of anaerobes. So an external observer looking for atmospheric disequilibrium as a sign of life would not have found it, even though Earth was 'alive'.

Evidence for extraterrestrial intelligence?

This chapter has focused mainly on evidence for the possibilities of microbial extraterrestrial life in our Solar System, and also considered exploration of our close stellar neighbourhood for Earth-like planets. There is an additional strand of exploration that I have not yet discussed: the Search for Extra-Terrestrial Intelligence (SETI). This is a very different line of enquiry, looking not for (non-conscious?) micro-organisms, but for advanced life forms capable of communicating with Earth. And so SETI uses telescopes to monitor the heavens for incoming signals. Why do this? Surely it is an impossible task to search, in effect, the whole of the Universe across all radiation frequencies every night, on the off-chance that an extraterrestrial civilization is trying out its long-range broadcasting abilities? Yes, indeed, that is an opinion held by many. But, argue the proponents of SETI, although it is a very slim chance, if we don't listen, then we will never hear, and it is therefore a chance worth taking. Of course, the search is not random – it is based on specific frequencies (related to the wavelength of the hydrogen molecule, the most abundant element in the cosmos), and only monitors 'likely' neighbourhoods, i.e. not the centres of galaxies where black holes lie, or supernova remnants, etc. The inspiration for SETI draws heavily on the ideas of Frank Drake, Emeritus Professor of Astronomy at the University of California at Santa Cruz, who, in the early 1960s, first put forward a mathematical expression of the probability of finding extraterrestrial civilizations. The 'Drake Equation' places factors such as stellar age, neighbourhood, etc., into a relationship, the answer to which is $n \geq 1$. We, on Earth, are the one civilization about which we know; SETI is looking to see whether $n > 1$. Results from (more mainstream) lines of research, such as the search for Earth-like planets, the distribution of the ages of stars in galaxies, the grouping of galaxies in clusters, etc., are fed into and inform the constraints used to direct SETI. It must only be a matter of time before a solution other than $n = 1$ is found for the Drake Equation. (Note: I don't specify how long the 'matter of time' might be!)

As a completely non-scientific, final, digression from the main subject material of this essay, I make a brief consideration of the chance that we have been visited by aliens from other worlds. Could we have been visited in the distant past, in the first billion years of Earth's existence, prior to the occupation of Earth's surface by complex organisms? I do not know, and I am not sure how we could tell. There have been lengthy volumes written explaining the origins of ancient historical sites (e.g. the pyramids in Egypt and Mexico, the statues of Easter Island, the Andean excavations pre-dating Incan civilization, etc.) in terms of blueprints and directions from visiting extraterrestrials, bent only on guiding the simple peoples of Earth. I am completely dismissive of such inter-pretations. Ancient civilizations were able to observe the sky and the motions of constellations and planets in a detail that is no longer available to us, now that we have obscured so much of the sky with light pollution. It is not sur-prising that they built monuments that aligned with the motions of the Sun, Moon and major planets. They could do this without guidance from the kindly, paternalistic extraterrestrials. There are also innumerable accounts of UFOs and visitors from space, some of whom are bent on abducting our citizenry for their own fell purposes. Again, I have no belief in such stories. Any sensible alien who wished to make contact with humanity would orbit Earth for a few revolutions, and then land in a populated region, not a backwater. I do not think that there is any evidence whatsoever that we have, as yet, been visited by an extraterrestrial civilization. What would I regard as evidence for such a visit? I think, at the very least, a very visual and obvious presence and a means of continuous two-way communication between their home-base and Earth. I hope, however, that I am sufficiently open-minded to accept evidence of extraterrestrial life when, eventually, it does make contact with us.

Summary: evidence for life beyond Earth?

So far, we have not yet found evidence for extant life beyond Earth; neither have we discovered evidence for extinct life beyond Earth. However, the discovery of micro-organisms on Earth that are able to survive in conditions of extreme heat, cold, pressure and radiation encourages the view that there may be life in habitable niches beyond Earth. There is an abundance of organic materials and water within the Solar System, and the potential for their presence on planets around other stars. But the evolutionary process on Earth has been affected by many external factors: 'The history of life is not necessarily progressive; it is certainly not predictable. The earth's creatures have evolved through a series of contingent and fortuitous events' (Gould, 1994). The 'fortuitous events' to which Gould refers include impact of asteroids – where would mammals be now if an impact had not wiped out the dominant species (dinosaurs) some 65 million years ago? But there are other, very special, features that Earth exhibits that are not duplicated on other planets within our Solar System. These include: the nature of Earth's structure (a molten core powering plate tectonics and generating a magnetic field); the presence of a single large satellite (Moon) stabilizing Earth's rotation; and a surface temperature at which liquid water is stable. So even if microbial life is widespread within the Solar System or Galaxy, it is unlikely that it would survive and evolve in the same way as has occurred on Earth. Nonetheless, we must keep on piecing together evidence for possible life beyond Earth, in the hope that one day either we will find it, or it will find us!

FURTHER READING

Gould, Stephen Jay, 'The Evolution of Life on Earth', *Scientific American* 274, 1994.

Grady, M., *The Search for Life*, London: The Natural History Museum, 2001.

Jakosky, B., *The Search for Life on Other Planets*, Cambridge: Cambridge University Press, 1998.

Lunine, J. L., *Earth: Evolution of a Habitable World*, Cambridge: Cambridge University Press, 1999.

Morris, S., *Life's Solution: Inevitable Humans in a Lonely Universe*, Cambridge: Cambridge University Press, 2003.

Rees, M., *Our Cosmic Habitat*, Princeton, NJ: Princeton University Press, 2003.

Ward, P. D. and Brownlee, D., *Rare Earth: Why Complex Life is Uncommon in the Universe*, New York: Copernicus, 2000.

5 Evidence in Theory: Superstrings and the Quest for Unification

BRIAN GREENE

Introduction

For a few thousand years, poets, philosophers, mathematicians and physicists such as myself have asked some of the most difficult questions: the kind of questions that cut to the core of what it means to be human, and to be part of the Universe in which we find ourselves. From the point of view of physics, these big questions include: What is space? Is space a real physical entity or is it just a useful but abstract idea? What is time? Does time have a beginning, does it have an end? Is there a smallest unit of time? What are the basic fundamental components making up the Universe and how do they interact and influence each other? How does that interaction drive the evolution of the cosmos?

Throughout history, these questions have been tackled by some of humanity's finest minds, but the most lasting contributions have been made by one person: Albert Einstein. Einstein dramatically changed our thinking about space, time and matter during the last century and much of what we now think about the Universe owes its origin to Einstein's insights. Even so, there was one particular goal that eluded even Einstein: finding what he called 'a unified theory'.

A unified theory would be able, in principle, to describe everything in the physical Universe – the astronomically large, the infinitesimally small and everything in between – using just one idea, one master equation. Einstein searched desperately for such a theory for the last thirty years of his life. Even on his death bed, in a Princeton, New Jersey, hospital in 1955, Einstein asked for the pad of paper on which he had been scribbling equations in a feverish hope that he might find this unified theory, that he might, in those final moments, conclude the arduous search that had driven him with such intensity for decades. He did not.

But this leads us to ask: What was it that drove Einstein with such unrelenting passion to find such a unified theory? Einstein himself writes eloquently of his inspirations:

> The most beautiful thing we can experience is the mysterious. It is the source of all true art and science. He to whom this emotion is a stranger, who can no longer pause to wonder and stand rapt in awe, is as good as dead: his eyes are closed.
>
> It is enough for me to contemplate the mystery of conscious life perpetuating itself through all eternity, to reflect upon the marvellous structure of the Universe, which we can dimly perceive, and to try humbly to comprehend even an infinitesimal part of the intelligence manifested in nature.

But that is only part of the motivation. The search for a unified theory has a more pragmatic imperative, as it turns out that there are fundamental questions about the nature of the Universe that can only ever be answered with a unified theory in hand. We'll begin with perhaps the biggest of the questions that fall into that category, a question about how the Universe began.

About 15 billion or so years ago, for some reason – we don't exactly know why – the Universe as we know it simply popped into existence in what we call the Big Bang. All the matter and energy there is in the large Universe we observe today – in fact all of space and time –was flung out from that fiery beginning. The Universe then expanded from this exceptionally small and dense beginning and, as it did so, it cooled. Material began to coalesce into structures that gave rise to the familiar features that we see today: stars, galaxies, clusters of galaxies. That, in a nutshell, is the history of our Universe. Of course, our current observations give us only the final few frames of this history. To run the 'cosmic film' back from today towards the beginning, we must rely on our mathematical understanding – the equations we believe underlie the evolution of the Universe – from which we can infer the nature of the Universe at earlier and earlier times, ultimately trying to figure out what started the story off in the first place.

When we do this, when we run the film backwards, we find that everything that is now rushing apart comes back together. Galaxies come closer and closer together, the Universe gets hotter and hotter, smaller and smaller, more and more dense. The cosmos implodes. Using our current theories, we can run the film back to a mere fraction of a second after the beginning, but if we try to go further the film dissolves into static. Noise. Garbage. Our current understanding of the laws of physics *breaks down* if we try to infer the nature of the Universe before about 10^{-43} seconds after the beginning. Even further back, and therefore time zero itself, remains cloaked in mystery. This, in my opinion and in the opinion of many other people who work in the field, is the major motivating question in the search for a unified theory: How did it all begin?

This chapter represents a status report on how far we have come in understanding what we believe might be the unified theory that eluded Einstein. It's a theory called 'Superstring Theory', or 'String Theory' for short. I will discuss the essential hurdles and the key achievements on the road to the construction of this theory and, in keeping with the theme of this volume, I will describe the nature of the evidence we use to try to persuade each other and the wider world that we are onto the right track, and what might, at a later date, be accepted as evidence that we have reached our goal.

Conflict: Newtonian gravity and Special Relativity

There are many points in the history of science from which one can begin to tell a story of the quest for a unified theory; we shall begin with the first quantitative attempt to understand gravity. As we will see, an understanding of gravity proves to be the critical element in the search for the deepest laws of the Universe. It was Isaac Newton, in the late 1600s, who took the first critical steps in understanding the influence of gravity and how to describe it mathematically. In doing so he formulated the famous Universal Law of Gravitation, that the force of gravity between *any* two objects is directly proportional to the product of the masses of the two objects and inversely proportional to the square of the distance between them, or in its rather more elegant mathematical expression:

$$F = G\frac{M_1 M_2}{r^2}.$$

In giving the world this equation, Newton provided us with a new and powerful path tool in searching for truth. Newton demonstrated – for reasons that we still don't understand – that mathematics provides the most insightful and economical language for investigating and understanding natural phenomena in the Universe. In short, this tiny equation underlying Newton's Universal Law of Gravitation really worked! With it, you can predict where the Moon will be a week from now, a year from now. You can foretell a solar eclipse 100 years from now, 1000 years from now. For that matter, you can work out from observations today when a solar eclipse took place 100 years ago, 1000 years ago. The evidence then that this first step encapsulates the force of gravity, allowing us to account for the motion of bodies through the Universe, is overwhelming. In fact, we still make use of this very equation when we are sending rockets to the Moon or to Mars; it is without question something in which people, especially scientists, have enormous faith. The mathematics appears to reflect reality. And so, it's not surprising that Newton's description of the Universe remained unchallenged for about 250 years.

But something dramatic happened at the turn of the twentieth century. A patent clerk, working in Switzerland, came to the shocking conclusion that, even though Newton's equation was doing such a great job of explaining data, it could not be right. There must be something wrong with it. Though it might be close to the truth, it could not be the whole story regarding the force of gravity. This patent clerk was, of course, Albert Einstein, and to understand how he came to this thoroughly heretical conclusion we have to go back to 1905, the year in which Einstein postulated his Special Theory of Relativity. One of its central statements, the second of the two fundamental postulates of the theory, is that nothing can go faster than the speed of light. And in that statement, the word 'nothing' means nothing! No signal, no disturbance, no influence nor information of any kind can travel from one point in the Universe to another at a speed greater than the speed of light. This speed limit is not something with which we are generally confronted in our everyday experience; the reason for this, simply put, is that light travels really fast. Light travels at 186 000 miles per second: fast enough to go around the world seven times in one second. But the fact that there is a speed limit at all spells the end of the hegemony of Newton's theory.

The reason is that, in Newton's theory, if you take the equations at face value, gravity exerts its influence from place to place in no time at all. Gravitational influence is propagated instantaneously – much faster than the speed of light. That's the problem. We can illustrate this with an example. Imagine it's a bright sunny day, you're outside walking and all of a sudden the Sun explodes. How long will it take before you realize that the Sun has exploded? The light from the explosion will have to travel all the way from the Sun to your eyes, a distance of around 93 million miles; at 186 000 miles per second, that takes about eight minutes. So, according to the postulate that nothing can travel faster than light, we cannot know that the Sun has exploded before eight minutes have elapsed after the instant of the explosion. However, according to Newton, it is the Sun that keeps the Earth in orbit. If the Sun goes away, our motion through space should change abruptly and we should feel it . . . instantly. According to Newton, then, we should feel the demise of the Sun through the change in our motion at the exact moment that the Sun explodes. Thus, we'll learn about this explosion by feeling it gravitationally before we learn about it by seeing it. We will learn about it gravitationally before light has had a chance to reach us. Gravity, therefore, travels faster than light. Herein lies the conflict.

Notice that this conflict was purely theoretical. There was no experimental data with which Newton's theory was incompatible. But the postulate that nothing could travel faster than the speed of light contradicted theoretical predictions of Newton's theory. Thus the two mathematical frameworks for understanding and interpreting data from the Universe were in conflict.

Resolution: Einstein's theory of gravity: General Relativity

This conflict set Einstein on a journey, a journey that would take him a decade, to try to invent a new framework for understanding gravity. His goal was not to interpret data that was incompatible with Newton's theory, though he did achieve that goal. His goal was rather to develop a theory of gravity that was not in conflict with his own Special Theory of Relativity. This turned out to be an arduous and sometimes tortuous task. In 1912, about five years into these endeavours, Einstein wrote to a friend that, compared with his current work, understanding the Special Theory of Relativity was mere child's play. That's how very difficult the problem of understanding gravity proved to be. Despite these tribulations, by 1915 Einstein had achieved his goal. He named the theory 'the General Theory of Relativity' and presented it to the world.

The General Theory of Relativity was a rather startling new way to think about gravity and, in fact, space and time. To get a sense of the magnitude of this upheaval, it is worthwhile to ask what Einstein was doing during that decade. Einstein was trying to figure out nothing less than the mechanism by which gravity operates. Although Newton had written down an equation that described with great accuracy the gravitational force acting between objects, he quite studiously avoided ever addressing the knotty issue of what it was

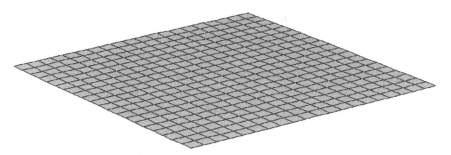

FIGURE 5.1. A schematic representation of flat space.

that constituted the mechanism by which this force acted – how the force was mediated. In his book *The Principia*, Newton says, in fact, 'To that important question [of the mechanism by which gravity operates], I leave it to the consideration of the reader.' Now, most readers would read this, shrug their shoulders and read on; but Einstein was not such a reader. Einstein was equal to the task of providing an answer to this most vexing question. And the answer he came up with, after a decade of struggle, is astonishing.

Einstein found that the mechanism by which gravity exerts influence is nothing less than the fabric of space and time. Before going into the details of this concept, we'll employ a familiar analogy to try and give a sense of the general idea. The analogy is one physicists love to use in this context. Putting space–time and gravity aside for a moment, imagine a large rubber sheet stretched out flat, as in Figure 5.1, and imagine a little marble rolling along the surface of this rubber sheet. If the sheet is completely flat, the marble will roll in a straight line. Now imagine that we place a large heavy object like a bowling ball in the middle of the sheet, causing the sheet to take on a curved shape. A marble rolling into this curved region of the rubber sheet will no longer travel in a straight line, but will travel in a curved path dictated by the curvature of the rubber sheet, as in Figure 5.2. This is the rough idea Einstein employed to understand gravity. In Einstein's General Relativity, the rubber sheet is replaced by the fabric of space–time. In the absence of massive objects, the fabric of space–time is flat and objects moving through space–time travel in straight lines. But massive objects, and indeed small ones, distort the fabric of space–time and in that way influence the trajectories of objects moving in their vicinity. A large object like the Sun will distort the fabric of space–time in a way very similar to that of the bowling ball on the rubber sheet. Closer

FIGURE 5.2. A massive body like the Sun causes the fabric of space to warp
somewhat like the effect of a bowling ball placed on a rubber sheet. The Earth is
kept in orbit around the sun because it rolls along a valley in the warped spatial
fabric. In more precise language, it follows a 'path of least resistance' in the distorted
region around the Sun, much as a marble would do travelling on a rubber sheet.

to home, the Earth distorts the space–time around it in a similar way, and
the Moon is kept in its orbit because it rolls around the trough in space–time
created by the Earth's presence somewhat – as in Figure 5.2 – as the marble
moves along a trough in a curved rubber sheet.

Space–time is thus the medium that transmits gravity – curved space–time *is*
gravity – according to Einstein. The mathematics that underlies this description
of gravity is based on a branch of multi-dimensional curved geometry called
Riemannian geometry. The key equation, Einstein's equation, has a disingenu-
ous simplicity:

$$R_{\mu\nu} - g_{\mu\nu}\frac{R}{2} = \frac{8\pi\, T_{\mu\nu} G}{c^4}.$$

Basically it says that the curvature of the Universe (the left-hand side of the
equation) is given by the amount of matter and energy that is present (the
right-hand side of the equation). Einstein's theory was a stunning development
because it filled in the missing piece of the puzzle. It provided a description of
the mechanism by which gravity operates. Gravity is caused by the curvature
of space and time. I feel the chair I'm sitting on pushing up on me, according
to Einstein, because my body wants to slide down an indentation in the space–
time continuum caused by the presence of the Earth and the chair is getting
in the way. That is gravity, according to Einstein's geometrical picture of how
the Universe works.

But Einstein's description of the mechanism of gravity is more than just a
pretty picture. When Einstein's equation is used to describe, say, the motion

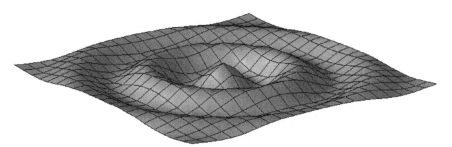

FIGURE 5.3. Gravitational waves – ripples in the fabric of space–time.

of the planet Mercury, the prediction matches the observational data with greater fidelity than the predictions made by Newton's theory. The orbit of Mercury was one of the first calculations Einstein made with his new theory. He later wrote that, when he found that the result of his theory resolved the discrepancy between observations and Newtonian theory, he was so thrilled it gave him heart palpitations. Since that time there have been many experiments that have confirmed that his theory does indeed describe the force of gravity – and the behaviour of space and time in the vicinity of massive bodies – with outstanding accuracy. But the motivation behind Einstein's odyssey to find a description of the mechanism of gravity was not to replace Newton's theory with one that better described experimental observation – notably the perihelion of Mercury. The motivation was the theoretical conflict between the apparent infinite speed at which gravity is mediated and the postulate that nothing should go faster than the speed of light. Which leaves the question of how Einstein's theory resolves this conflict?

To answer this, consider once again the rubber sheet, now free of bowling balls and marbles. Tapping the rubber sheet causes ripple-like disturbances to spread out away from where you tap, much like the way ripples travel outwards when you throw a pebble in a pond. Einstein realized that the same thing happens with the fabric of space–time. The ripples mediate gravity, so if you can work out how fast the ripples travel through space – how fast fluctuations in the curvature of space travel – then you will have worked out the speed at which gravity exerts its influence. Einstein made this calculation and found that, far from travelling instantaneously, these gravitational disturbances travel at a very particular speed. And that speed turns out to be exactly the speed of light. Einstein's new theory thus provided a description of gravity entirely consistent with the second postulate of his earlier Special Theory of Relativity. When the Sun explodes, it sends out something like a tidal wave in space. During those eight minutes it takes light to travel from the Sun to Earth, the space–time in the vicinity of Earth is unaffected and Earth travels along as before, oblivious of the incoming gravitational disturbances. Only when those disturbances have reached Earth can Earth feel any gravitational effect. This implies that we see the Sun disappear at exactly the same moment as we feel it. Conflict resolved. Gravity and light travel at the same speed. The mathematics of gravity and the Special Theory of Relativity are brought together in a harmonious whole.

The development of the General Theory of Relativity was a giant step forward in the development of our understanding of the nature of the physical Universe. Gravity was brought into harmony with the central postulates of Special Relativity, and the development fundamentally changed the way we think about space and time. But as is often the way with these things, just as soon as you solve one problem, another crops up, and that's exactly what happened here. The same kind of conflict that motivated the development of General Relativity came back to bite it, once it was born. This is because the mathematics of General Relativity conflicts with that of Quantum Mechanics.

Quantum Mechanics

Quantum Mechanics was developed in the 1920s and 1930s by scientists who recognized that, when they tried to apply the then-current theories – essentially those of Newton and Maxwell – to molecules and atoms, they found that, although the laws worked well on large everyday scales, they failed miserably when applied to molecules, atoms and subatomic particles. For example, those theories applied to a single atom would predict that every single atom in the Universe should self-destruct in a fraction of a second. That, fortunately, doesn't happen. It was clear that a new theory applicable to atoms and subatomic particles was necessary, and the theory that the generation of physicists in the 1920s and 1930s came up with to address these problems is called 'Quantum Mechanics'.

Quantum Mechanics is a strange and counter-intuitive discipline. It describes the mechanics of atoms and subatomic particles in a language and using concepts that are entirely alien to our everyday experience. The theory speaks of objects having both a particle-like and a wave-like existence; it speaks, in one interpretation, of the possibility of parallel universes; it allows for something called 'Quantum Tunnelling'. The latter concerns the notion that, if I were to try over and over again to walk through the wall of my office into the office of my colleague next door, I would find that I would just bang into the wall and bounce off. But Quantum Theory says that there is a small chance that on one of those attempts I'll actually tunnel right through the wall and emerge unscathed on the other side; it's a *small* chance, but it's not a zero chance. And in the microscopic version of exactly that experiment, where an electron is fired at a barrier which the theories of Newton and Maxwell would say there is absolutely no chance that the electron can penetrate, the Quantum Theory says that there is a small chance that it will, and, indeed, when the experiment is performed, every so often it does.

Quantum Mechanics, like General Relativity, is a phenomenally accurate theory. Calculations that have been done over the last 20 or 30 years have shown that Quantum Mechanics can explain experimental data with unprecedented precision. For example, Tom Kinoshita at Cornell University has spent decades calculating the magnetic properties of particles like electrons – the calculations run to thousands of pages. At the end of these many many pages of calculation, there is a number, a number that describes the magnetic properties of the electron. When this number, predicted from Quantum Mechanics, is compared with measurements of the property this number represents, the results agree to the tenth decimal place: *one part in 10 billion*. The evidence for the validity of Quantum Theory is thus overwhelming, but again there is a problem. The problem is that the structure of Quantum Mechanics conflicts with Einstein's theory of gravity, and the conflict arises from an aspect of Quantum Mechanics called Uncertainty.

A key element of Quantum Theory is a relation known as the Uncertainty Principle that was developed in 1927 by Werner Heisenberg. The Uncertainty Principle works similarly to the special-order menu of my local Chinese restaurant in Manhattan's Chinatown. On this menu, there are two columns, column A and column B, and if you order the first dish in column A then you're not allowed the corresponding dish in column B and vice versa. That, in a nutshell, is the Uncertainty Principle. To be a little more precise, the Uncertainty Principle says that knowledge of the microscopic world is very much like that menu divided into two columns: knowledge of items in the first column fundamentally compromises your ability to know about the corresponding items from the second list and vice versa. For example, at the top of menu A we find 'position' and the corresponding entry on menu B is 'velocity'. If you know where, say, an electron is, Quantum Uncertainty says that that fundamentally compromises your ability to know how fast it is moving. Likewise, if you know how fast it's moving, that fundamentally compromises your ability to know where, exactly, it is. This 'ability to know' has nothing whatsoever to do with the equipment you use or how good an experimenter you are. It is a fundamental limitation on our knowledge of the physical world. This is very non-intuitive. Our everyday experience tells us that not only can I say that my desk is right here, I also know exactly how fast it's moving (it's not moving at all), and then I've specified both its position and its velocity. But this is misleading, because if I were to examine the microscopic components of my macroscopic desk, I could not simultaneously nail down both their speed and their position with complete precision. This is the uncertainty dictated by Heisenberg's principle.

The consequence of Uncertainty that leads to conflict with General Relativity is the observation that, because we cannot nail down the position and speed of objects in the micro-world, the micro-world is jittery, turbulent, chaotic, frenzied. Objects are free to fluctuate among all manner of possible states, consistent with our limited knowledge of the physical world, as dictated by Heisenberg's principle. It is this jittery quality of Quantum Mechanics that sets it in conflict with Einstein's General Theory of Relativity.

Conflict: General Relativity and Quantum Mechanics

General Relativity describes the Universe as a gently curving geometrical object. Though curved, the curvature is not severe, it's smooth and gentle. Quantum Theory, on the other hand, paints a picture of a world that is anything but gentle. It's the jitteriness of Quantum Theory versus the gentle character of General Relativity that entails a conflict between them. More precisely, this manifests itself whenever you try to put Quantum Mechanics and General Relativity together to do a calculation. Almost invariably when you do this, you find your calculation yields one and the same answer: infinity. More precisely yet, General Relativity, in the most naïve attempts to merge it with Quantum Mechanics, proves to be a non-renormalizable theory. As anyone who has ever studied Quantum Field Theory can tell you, infinities crop up in our description of the other forces of Nature, but there are techniques available for circumventing these infinities – the infinities can be absorbed into the physical parameters we use to describe the system we are studying: the mass of the electron, for example, or the electric charge on the electron. But if we try to apply the same techniques to quantize General Relativity, it simply does not work.

We can visualize this conflict if we imagine looking at the fabric of space–time at ever smaller distances, starting on everyday scales and looking closer and closer at the structure of space and time and matter. We can imagine going smaller than the scales of molecules and atoms, and then going yet further down to the scales of subatomic particles: protons, neutrons, electrons and so on. If we keep on going, to a scale 100 billion billion times smaller than the scale of an atomic nucleus, then it is here, on a scale of about 10^{-33} cm, that the conflict between Quantum Theory and General Relativity can be seen. Space–time on such tiny distances is no longer smooth, as it was in Einstein's picture. Instead, Quantum Uncertainty gives space–time a frothing, undulating quality, as in Figure 5.4. In short, the two theories lock horns. Notice, though, that if we now travel back out again to larger scales, this turbulent, frothy quantity becomes less and less violent. There is always a certain amount of uncertainty, but it is small enough that it remains under control and, the larger we get, the smaller the uncertainty becomes. Gradually as we look to larger and larger scales, the Universe begins once again to take on the gentle placid geometrical form on which Einstein based his theory of General Relativity. At these scales, the equation of General Relativity works extremely well, but in the ultra-microscopic scales of around 10^{-33} cm, the picture is spoiled by Quantum Uncertainty, or, technically speaking, by the non-normalizability of Quantum Gravity.

Once again, we are confronted with a conflict, and as was also the case for the conflict between Newtonian gravity and Special Relativity, it is not a conflict between theory and experiment. General Relativity works. Quantum Theory works. Both theories work spectacularly well at explaining experimental observations. The conflict is theoretical and only arises at incredibly tiny scales, 100-billion-billionth of the size of an atomic nucleus, scales far below what we can probe experimentally. You might well ask, why are we interested in a conflict that only manifests itself at such infinitesimal scales? We have theories for everything we might expect to have to deal with in the physical world. Why not stop there and call it a day?

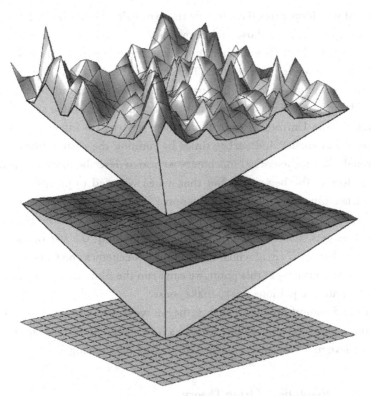

FIGURE 5.4. By sequentially magnifying a region of space, its ultramicroscopic properties can be probed. Attempts to merge General Relativity and Quantum Mechanics run up against the violent quantum foam emerging at the highest level of magnification.

There are a number of reasons why, despite the overwhelming success of the extant theories, people have been devoting their professional lives to resolving this theoretical conflict. The first resonates with the theme of this volume. We see a mathematical inconsistency – this disintegration of theory into nonsense when it is applied at very small scales – as powerful evidence that our theories are incomplete, that there is a deeper understanding to be attained. Regardless of how extreme and tiny the scales are to which the theory must be pushed in order for it to break down, the fact that it breaks down at all is clear evidence that we must continue to strive to find a more complete theory. We simply cannot accept that the Universe is described by two mathematical structures, each of which operates flawlessly in its own realm, but each of which conflicts with the other where those realms overlap.

In any case, there is a further, more practical – at least by these standards – reason for resolving this conflict. The conflict means that in the extreme and unusual situations in which General Relativity and Quantum Mechanics must both be used, we have no consistent theory, and one of these situations is the origin of the Universe. Recall that, when we tried to infer the state of the Universe at earlier and earlier times by running the cosmic film backwards through the mathematical machinery we use to describe it, we reached a time very close to the beginning. But that, when we tried to push the clock back to earlier times, our theories broke down and all we obtained was nonsense. We are now in a position to see why this is so. As we run the film back the Universe gets smaller and smaller and at some point the size of the Universe reaches that 10^{-33} cm at which the Quantum Mechanics and General Relativity come into conflict – at this point, we can't run the film any further back as our mathematical equations fail to make sense.

To understand how the Universe began we're going to have to resolve this conflict. Since the mid-1980s, scientists have come up with an approach that we believe may resolve the conflict. It's called 'Superstring Theory'.

Resolution? String Theory

String Theory tries to answer a question that mankind has posed for thousands of years: What are the fundamental building blocks of the Universe? More precisely, what are the fundamental building blocks of matter and energy? If we were to take a piece of wood and cut it in half, cut it in half again and cut it into ever smaller and smaller pieces, where would it end? We all know, of course, that at some point we get to atoms. But atoms are not the end of the line, because we know that atoms can be split. Electrons orbit around a nucleus made up of neutrons and protons. Neutrons and protons, we learned in the late 1960s, are themselves made up of further constituents known as quarks. Electrons and quarks, though, is where traditional theory and current experiment have stopped. To interpret the data we receive from particle accelerators, we need only the language of electrons, quarks and a handful of other exotic species of particles. That's it. These, for us, are the un-cuttable dots out of which matter and energy are built; fundamental dots with no internal structure.

String Theory challenges that picture. String Theory suggests that there is another layer of structure. Inside an electron, inside a quark, inside any particle you care to name, is something else. This something else is a little string-like filament of energy that can vibrate in different patterns like the string on a violin or a piano. Now, when the strings on a violin or a piano vibrate in different patterns, we hear them as different musical tones. Rather than producing music, the different vibratory patterns on the strings in String Theory correspond to different particles. More precisely, the resonant vibrational patterns give rise to the familiar characteristics of mass, charge and the nuclear properties that we associate with particles. That, in essence, is String Theory. Everything around us, at its most microscopic level, would resemble a wealth of little tiny vibrating strings, vibrating matter and energy into existence through the different patterns of vibration that they can produce.

That, briefly, is what the theory tells us about the nature of matter and energy. And what this theory does for us is rather spectacular. The world-view of String Theory allows us to resolve the conflict between General Relativity and Quantum Mechanics. This is why this theory has attracted so much interest and why it is taken seriously. We believe that this theory provides the key to unlocking the door to the Einsteinian unified theory – it provides a single unified framework that may be able to describe everything, absolutely everything, in the Universe.

But how does String Theory resolve the conflict between General Relativity and Quantum Mechanics? Where, in the old world-view, we had point particles, in String Theory these are now replaced by strings extended, albeit not very far, in space. This has the effect of smearing out the point particle into a loop. Not only does this smear out the particle, it also smears out those violent jitters in space that are the origin of the conflict between Quantum Theory and General Relativity. Space still has this undulating character, but the undulations are not as violent as they were. They are spread out and hence diluted by the passage from point particles to strings, ameliorating the problem. In essence, String Theory quiets the ferocity of Quantum Uncertainty just enough for General Relativity and Quantum Theory to fit perfectly together.

More precisely put, as anyone who has studied Quantum Mechanics can tell you, not only does Heisenberg's Uncertainty Principle declare uncertainty, it declares with complete certainty the amount of uncertainty there is on a given length scale. What String Theory effectively manages is to set a fundamental *minimum* length. The String is the fundamental entity, and whereas previously our fundamental entities were infinitesimal points, the String has a length associated with it. If you use a string to probe an object in a scattering experiment, there is no way to probe scales smaller than the string itself. Thus, there is a fundamental limit to how small you can go; in the language of the field, there is an ultraviolet cut-off. This short-distance cut-off is exactly such as to remove the infinities that would have been encountered in a more traditional calculation. This is how String Theory gives us a Quantum Theory of gravity, which doesn't suffer from the infinities of all previous approaches.

Some consequences of String Theory

The implications of String Theory for our understanding of the nature of space and time are many and bizarre. As I've emphasized earlier, the mathematics works. But what I have yet to emphasize is that, more precisely, the mathematics only works if our world has more than three spatial dimensions. More than left–right, up–down and back–forth. In fact, for this theory to hang together mathematically, our Universe should have at least six and probably seven extra spatial dimensions beyond the three with which we are so intuitively familiar. Now you might well argue that this is clearly absurd and should surely represent quite categorical evidence that String Theory is wrong since it makes predictions that clearly contradict observation. However, this is not the case, and indeed my career has focused to a large extent on the issue of the extra dimensions. My colleagues and I have come up with a number of ways by which this seemingly ridiculous requirement can be made to correspond to our experience of our three-dimensional reality. I'll focus on just one here.

Dimensions, we believe, can come in two sorts: big and obvious like the ones we see around us, or tiny, curled up and virtually impossible to detect. We suppose that we see only three because only three are of the big sort while the remaining are tiny and curled up; they're all around us, but too small not only for our eyes but also for our equipment to detect. To see how this can work, imagine a cable stretched over a canyon, and viewed from a distant vantage point.

From far away, the cable looks one-dimensional; it appears to only have length as you can't really see its thickness from such a distance. But if you imagine travelling down to smaller scales, going in close and adopting, say, the

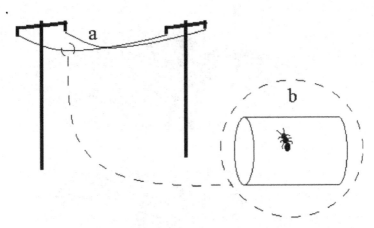

FIGURE 5.5. a) A power cable viewed from a substantial distance looks like a one-dimensional object. b) When magnified, a second dimension in the shape of a circle and curled around the cable becomes apparent.

perspective of a small ant, you realize that not only does the cable have the obvious dimension along its length, but it also has an extra dimension, the circular dimension that wraps around its girth (see Figure 5.5). From the perspective of large scales, this tiny circular dimension is difficult to see, but nevertheless it exists.

We think the extra dimensions String Theory requires of our Universe might have a similar nature. The three familiar dimensions are like the horizontal extension of the cable. The extra dimensions are like the small circular dimension curled around the cable – only visible close up or via tremendous magnification – and in the case of the extra dimensions of String Theory, you'd need *tremendous* magnification.

Figure 5.6 gives an idea of how this analogy might correspond to the Universe. It only shows magnification of two of the three familiar spatial dimensions, but when we approach the tiny scales at which these extra dimensions begin to manifest themselves we can see an extra dimension: not only can we move back and forth and left and right along the flat two-dimensional plane of the two large dimensions, we can also take a little tour around one of the tiny curled-up dimensions represented by the small circles on the picture (see Figure 5.7).

To know where we are, we need to know not only at what point we are on the two-dimensional plane, we also need to know how far around the circle on that point we are. Here we illustrate two large dimensions and one curled

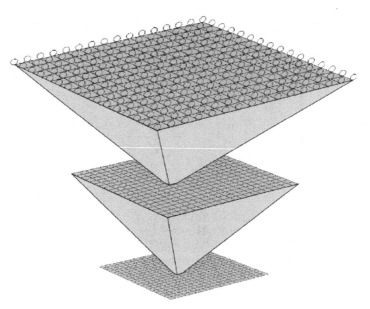

FIGURE 5.6. As in the schematic figure (corresponding to 5.1), each subsequent level represents a huge magnification of the spatial fabric displayed in the previous level. Our Universe may have extra dimensions – as we see by the fourth level of magnification – so long as they are curled up into a space small enough to have as yet evaded direct detection.

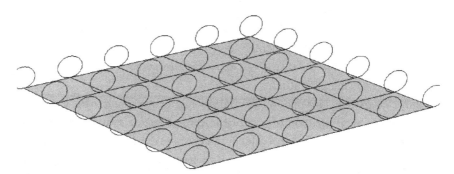

FIGURE 5.7. The grid lines represent the extended dimensions of common experience, whereas the circles are a new, tiny curled-up dimension. Like the circular loops of thread making up the pile of the carpet, the circles exist at every point in the familiar extended dimensions – but for visual clarity we draw them as spread out on intersecting grid lines.

FIGURE 5.8. A representation of a Calabi–Yau space.

up dimension, but our Universe has three large dimensions and String Theory suggests that there are six or perhaps seven extra curled-up dimensions.

String Theory not only tells us how many extra dimensions there are, it also tells us something about their shape. In the cable analogy and Figure 5.7 we illustrated the extra dimension by using little circles, but the shapes of the extra dimensions predicted by String Theory are considerably more complex. These shapes are known as 'Calabi–Yau shapes' or 'Calabi–Yau manifolds'. Figure 5.8 shows a representation of such a surface.

Mathematically, they are not quite as complicated as they might at first seem visually. In the language of differential geometry, they are a Ricci-flat complex, Kahler six-dimensional manifolds (three complex dimensions). Calabi–Yau manifolds are a field of study in themselves, but from a purely illustrative point of view, if String Theory is correct, then on these tiny scales the Universe looks something like Figure 5.9.

The evidence for String Theory

Make no mistake, I am totally forthright in declaring that there is absolutely no experimental evidence whatsoever that String Theory gives a valid description

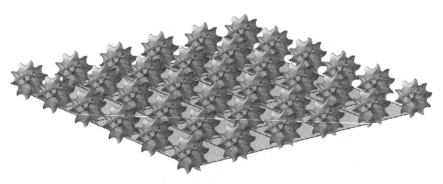

FIGURE 5.9. Two of the three familiar spatial dimensions with the hidden dimensions in their Calabi–Yau form.

of the Universe in which we live. You may ask: Why, then, do my colleagues and I persist in this pursuit? Is our faith in the mathematical consistency between Quantum Theory and General Relativity sufficient evidence? It certainly is not. Instead, we rightly view String Theory as a work in progress. We are attempting nothing less than a unification of the fundamental laws of the entire Universe, on all scales, from the unimaginably small 10^{-33} cm at which the conflict between General Relativity and Quantum Theory emerges, to the equally unimaginably large size of the known observable Universe, and everything in between. We've looked at this particular approach for 25–30 years. This is a pitifully short space of time for a potential achievement of that grandeur.

If we fundamentally believed that this theory would never make contact with direct experimental observation, if we thought it could never be subject to the scrutiny of empirical validation, we would drop it immediately. For the present, mathematical consistency must suffice, but we are confident that, as we proceed and acquire a deeper understanding of this theory, contact with experiment will be forthcoming. We shall transcend the mere mathematical evidence and finally connect with the real world.

String Theory suggests, for example, that there might be certain particle species, supersymmetric particles, that might be found by the Large Hadron Collider when it comes into operation in Geneva in 2008. We cannot, unfortunately, say exactly what these particles would look like; we can't yet determine from our current (incomplete) mathematical understanding of String Theory the masses of these particles. And that is a problem since it means that String Theory is not yet a falsifiable theory, because, for example, our failure to find the supersymmetric particles wouldn't prove String Theory wrong. It might just reflect that they are less massive than the Large Hadron Collider (LAC) is able to access. Neither, in fact, will it prove String Theory categorically correct should we find them, because versions of supersymmetric particles arise in other, non-String-based theories (although it is worth pointing out that super-symmetry was discovered in the context of String Theory and is an essential part of String Theory's mathematical structure). Instead, the discovery of supersymmetric particles would amount to circumstantial evidence for String Theory.

On another track, it is also possible that we might indirectly observe the effects of the presence of the extra dimensions predicted by String Theory: we have recently realized that energy in the high-energy collisions at the LHC might seep out from our familiar, large dimensions and into these extra dimensions. This would show up by having less energy at the end of various particle collisions than we had at the beginning. If such energy-loss events are observed, this would give us strong, indirect evidence for the reality of the extra dimension and hence for a key aspect of String Theory.

But there are other approaches to grounding this highly esoteric and abstract subject in reality. Perhaps the grandest, and the one that would represent the most persuasive evidence for the validity of String Theory as a unifying Quantum Theory of gravity, is the following.

Over the last century, experimenters have measured to fantastic accuracy twenty or so quantities that are necessary for prediction using our current theories. These are quantities like the mass of the electron, the mass of the quarks, the strength of gravity, the strength of the electromagnetic force and so on. These quantities have been measured and we know their numerical values, but we do not know why they have the very particular values that they do. It turns out that the actual values these numbers have are extremely significant, because if these numbers differed by just a few per cent from their measured values, if the mass of the electron were just a little lighter, or the strength of the electromagnetic force were just a few per cent stronger, then the Universe as we know it could not exist. Nuclear processes could not happen, stars would not shine, planets would not form, and life could not exist. Why is it, then, that these numbers have exactly the 'right' values for the Universe as we know it to exist?

Our current theories have no answer. String Theory has no answer either, but String Theory at least provides a framework in which the question may be sensibly posed and from which we might realistically suppose an answer may one day emerge. The framework is based on the observation that, in String Theory, these numbers are determined by the vibrational patterns of the strings. These strings are so small that they vibrate in all the ten or eleven dimensions String Theory requires – the three familiar spatial dimensions and the six or seven further, hitherto hidden dimensions curled up into various geometrical manifolds. The precise form of these manifolds affects the ways in which the string is able to vibrate, much as the precise form of a French horn affects the vibrational patterns of air streams rushing through it. Thus, the geometry of the extra dimensions plays a critical role in the allowed patterns of string vibrations and hence informs the relationships between the twenty numbers. If we knew exactly what these extra dimensions looked like – we don't yet – we ought to be able to calculate how the strings vibrate and in that way predict these twenty or so numbers. If the results of such a calculation agreed with the measured data, that would be overwhelming vindication of String Theory and the whole framework of 10- or 11-dimensional space. If we could achieve this, then for the first time in the history of science we would have a theory that not only describes the evolution of the cosmos but also explains why the Universe is as it is. That is our goal.

6 Statistics and the Law

PHILIP DAWID

The disciplines of Statistics and Law seem light-years apart, their spheres of interest disjoint and their practitioners inhabitants of different planets. Certainly for much of my own professional life as an academic statistician I gave no thought whatsoever to legal issues, until William Twining, Professor of Jurisprudence at University College London, opened my eyes to the fact – obvious in retrospect – that both disciplines share the same fundamental concern: making sense of evidence. Our ensuing interaction has resulted in my involvement, both as an academic and as an expert witness, in a number of issues arising at the interface between Statistics and Law. It turned out, very much to my own initial surprise, that many of these are intellectually fascinating and challenging, as well as vitally important for the fair administration of justice.

In this chapter I first discuss in some detail a number of statistical and logical issues arising from recent high-profile cases involving multiple infant deaths. The issues are subtle, and common sense a lamentably poor guide. I then address similar features of forensic DNA identification, before going on to describe some formal tools, based on graphical representations, that have been found useful for structuring and handling evidence in complex cases. These also offer great promise for application to many other fields of enquiry.

Sally Clark

The case of Sally Clark is the most celebrated of a number of recent cases in the UK in which mothers have been accused of murdering their babies, the evidence against them being wholly circumstantial. Cherie Booth also discusses the

This work was supported by the Leverhulme Trust and the ESRC. I am grateful to Terence Anderson, Ian Evett, Richard Leary, Julia Mortera, David Schum and William Twining for their invaluable past, present and future inputs and collaborations.

case elsewhere in this volume (see Chapter 7). It is worth considering at some length, because influential views on both sides of the case have been based on dubious statistical arguments.

One evening in December 1996, Sally was alone in her house with her first-born son Christopher, aged $2\frac{1}{2}$ months, who until then had seemed a healthy child. Two hours after his early evening feed his mother found him apparently dead in his bouncy chair in his parents' bedroom, and called an ambulance. Resuscitation was attempted but was unsuccessful. At post-mortem the pathologist found signs of bruising to the legs and damage to the frænulum inside the mouth, as well as some abnormalities in the lungs. His conclusion was that death was due to a lower respiratory tract infection.

In January 1998, Sally's second child Harry died aged 2 months in almost identical circumstances, both parents being in the house this time. Because of Christopher's previous unexpected death, Harry had been subject to intensive medical monitoring. Again, he had seemed in good health right up to his death. The post-mortem examination found signs of recent bleeding at the back of the eyes and in the spinal cord. Alerted by the pathologist, the police initially suspected both parents of actively bringing about their children's deaths, but soon focused on Sally alone. She was accused of having murdered her two children by smothering.

At committal and trial, the principal evidence for the prosecution came from medical experts. One of these was Sir Roy Meadow, Professor of Paediatrics and Child Health at Leeds University, who had examined the medical evidence. He testified on a number of medical issues, in which he could reasonably profess some expertise; and also on some statistical issues, in which he could not (in the words of the mathematician and cosmologist Hermann Bondi, 'Unhappily, the understanding that Statistics is a difficult subject is not widespread, even among distinguished paediatricians'). The main thrust of his statistical evidence was the extreme rarity of two infant deaths occurring from unexplained natural causes (Sudden Infant Death Syndrome, or 'SIDS' – otherwise known as 'cot death') in a family such as the Clarks. Using figures from an epidemiological study of the incidence of SIDS, he claimed that the overall incidence of SIDS was 1 in 1300 births, falling to about 1 in 8500 if one took into account various characteristics of the Clark family (that they were nonsmokers, had waged income, and the mother was over 26). He further testified that the chance of a repeat occurrence, once a first SIDS death has happened, would be essentially the same as for the first. That would imply that the probability of two SIDS deaths, in a family like the Clarks, could be calculated by multiplying 1 in 8500 by itself, leading to an overall rate for double SIDS deaths of about 1 in 73 million in such families.

There was no witness at the trial with any qualification in Statistics, and no serious cross-examination of the above argument. Summing up, Mr Justice Harrison said: 'However telling you may find those statistics to be, we do not convict people in these courts on statistics', but 'it may be part of the evidence to which you attach some significance'. On 9 November 1999, the jury at Chester Crown Court found Sally Clark guilty, by a 10–2 majority, of smothering her two babies, and she was sentenced to prison for life. The press immediately seized on the figure of '1 in 73 million' as incontrovertible evidence that Sally Clark was a wicked woman who thoroughly deserved to be locked away.

In January 2000, the *British Medical Journal* published an editorial, 'Conviction by mathematical error?' by Stephen Watkins, an epidemiologist. He claimed that Meadow's presentation of the figure of 1 in 73 million was based on a serious misunderstanding of probability theory. While, as discussed below, this may well have been so, Watkins' analysis was also fundamentally flawed. He argued that we should entirely disregard the first death (Christopher's), because this event had already been taken into account in drawing attention to Sally Clark in the first place – a principle that, if accepted more generally, would make it impossible ever to convict in any case where there was no evidence beyond that which led to arrest. In any case it was the second death (Harry's) that had aroused suspicion of foul play.

The effect of Watkins' recommendation would be to replace the squared figure, 1 in 73 million, by the much larger (and hence less incriminating) single-death figure, 1 in 8500 – or even, according to some studies of the rate of a second SIDS death, 1 in 1700. Meadow's eventual response (in January 2002) was that the statistical argument was never anything but a minor diversion from the medical evidence. Meanwhile, there had been a growing chorus of unease about Sally Clark's conviction, and a flurry of newspaper articles and radio and TV programmes arguing her innocence. Many of these again homed in on the 1 in 73 million statistic, but now aiming, like Watkins, to discredit it, or replace it with a larger and less incriminating figure.

The statistical issues

My own involvement in the case began when I was asked to contribute to the *Dispatches* programme about the case on Channel 4 television on 27 April 2000, and was briefed on the background. Later (October 2000) I was engaged as an expert statistical witness for the defence at Sally's appeal hearing. Both in the TV programme and in my expert report I made two main points. The former has been widely echoed, but the latter, which is by far the more significant, much less so.

Point 1. Independence? To calculate the probability of two SIDS deaths in Sally's family, we must not, without further justification, simply square the rate for a single death, as Meadow did. We can do so when we can argue convincingly for the *independence* of those two deaths: i.e. that, after identifying the appropriate SIDS rate to apply to the death of Christopher, the same figure would apply to the death of Harry, even after taking the fact of Christopher's death into account. But, on purely common-sense grounds, this is implausible. Even after taking into consideration the specific features (nonsmokers, etc.) of Sally's family used to home in on the figure of 1 in 8500, the two children must have shared many further characteristics, both known and unknown – most obviously, shared genes and domestic environment. The very fact of Christopher's death then gives some reason to believe that there might have been some potentiating factor, which could affect both brothers; and this in turn would increase the chance that Harry, too, would be affected. (As already mentioned, there is in any case some epidemiological evidence that death by SIDS is much commoner after a previous SIDS death in the family.)

Many similar criticisms of Meadow's independence assumption have been aired, to the discredit of the infamous 1 in 73 million statistic. In a widely quoted sound bite in a BBC Radio 5 documentary broadcast in July 2000, Peter Donnelly, Professor of Statistical Science at Oxford, said 'Unless the independence has been established, it's wrong. In that sense it's not rigorous, it's just wrong.'

Point 2. What should we be looking at anyway? Suppose, however, we could all agree on a figure for the probability of two deaths by SIDS in a family such as Sally's. Why should this number, of itself, automatically be regarded as interesting and relevant?

An obvious response to this question is that a rare event can reasonably be supposed not to have happened – all the more so, the smaller its initial probability. If we accept this reasoning, the tiny probability for naturally occurring deaths in this case constitutes strong evidence for the alternative hypothesis of Sally's guilt. We can plausibly surmise that the trial jury, like the press, could have been open to and swayed by such an interpretation of Meadow's statistical evidence, if only subconsciously – particularly in the light of the impressive-sounding '1 in 73 million' figure.

A more quantitative version of this qualitative reasoning has been dubbed 'the prosecutor's fallacy'. This involves regarding the figure of 1 in 73 million – actually a measure of the initial rarity of the event 'two SIDS deaths' – as the appropriate measure of the probability that that event has happened in this case. This reasoning would rate the probability that Sally is innocent at an entirely negligible 1 in 73 million, so providing overwhelming proof of her guilt. It is clear why such an argument would appeal to prosecutors! There are abundant examples of this reasonable-sounding but in fact totally fallacious argument being accepted and applied, often implicitly and indeed unconsciously, by judges, juries, journalists and the man on the Clapham omnibus – and of consequent miscarriages of justice.

But even in its weaker qualitative form, the argument, while superficially appealing, is just plain wrong. For we could use it to deduce that Christopher and Harry are still alive – since the initial probability of their both dying of any cause at all was surely very small. The evident absurdity of this conclusion demonstrates the illogicality of the reasoning.

So what to do?

At trial it was already known that the very rare event of two infant deaths had happened – that was never in question. Rather, what was at issue was to decide between two possible versions of this event, both initially extremely rare: had Sally's children died of natural, or of unnatural, causes? In particular, as observed by Cherie Booth, from this perspective, we should immediately appreciate that the probability that both babies would be murdered, which had never even been considered by the court, must have at least as much relevance to the case as did Meadow's probability that they would both die of SIDS. Applying Meadow's own multiplication approach to relevant official statistics yields a figure for the probability that two babies in one family will both be murdered of about 1 in 2 billion. Although this specific calculation is at least as spurious as Meadow's, it is merely proffered as an illustration of what needs to be considered.

Now whichever of these two competing versions we consider (both babies die of SIDS, or both die of murder), its initial probability is certainly very small indeed – and both these tiny probabilities are equally relevant to deciding the case. While one can readily envisage prosecution and defence brandishing their respective tiny statistics in adversarial combat, it is important to note that, from the statistical viewpoint, there is a unique correct way of proceeding, which involves combining, rather than contrasting, these two numbers. And when this is done, it turns out, surprisingly, that the small absolute values of these probabilities are simply not, after all, of any intrinsic interest or importance. Rather, it is the ratio of these probabilities – the relative odds for comparing the two alternative stories – that alone carries the essential information needed to choose between them.

Using fictitious figures to illustrate the form of the argument, suppose that, for a family like the Clarks, the probability of two infant deaths from SIDS is taken to be 1 in 5 million, and that for two infant murders to be 1 in 15 million. The all-important *ratio* of these probabilities is 3; double infant death by SIDS is three times more likely than double infant death by murder. Now in Sally's case we have in fact observed a double death. Supposing we can exclude any other possible causes apart from SIDS and murder – and ignoring all other evidence in the case – we now know that one of these must be the cause of the observed event, so that their probabilities add to 1. So, to respect the previously calculated odds of 3:1 between the two causes, the probability that the babies both died from SIDS must be three quarters, and the probability that they were both murdered is one quarter.

To elaborate on the above reasoning, consider a hypothetical population containing (say) 150 million families, essentially identical with that of Sally Clark. Out of these we would expect to find about 30 in all (1 in 5 million) in which both babies died of SIDS; and about 10 families (1 in 15 million) in which both babies were murdered – a total of 40 families in which two babies died. We know that, in Sally Clark's family, both babies died: it is one of these 40 families. Since the infant deaths were due to SIDS in 30 of the 40 families, and since, in the absence of any further evidence, we have no reason to consider Sally's family as different in any relevant way from the other 39, the probability that Sally Clark's babies died of SIDS is 30/40, i.e. three quarters;

and correspondingly the probability that they were murdered is 10/40, i.e. one quarter.

Applying the above logic to the pair of figures (both admittedly highly suspect) in the actual case – 1 in 73 million for two SIDS deaths and 1 in 2 billion for two murder deaths – their ratio (1/2 billion)/(1/73 million) = 0.0365 would give the odds on Sally's guilt given the evidence of the two deaths. This corresponds to a guilt probability of only 3.5 per cent. Note how different this conclusion is from that resulting from the prosecutor's fallacy, which would put the probability of guilt at 1 minus (1 in 73 million), i.e. as close to 1 as makes no difference. In contrast to such seemingly overwhelming evidence, the conclusion of the correct analysis would certainly not be enough to dispel 'reasonable doubt'. Even though truly appropriate figures in such a case may be hard to specify or agree on, the general thrust of the correct argument, combined with 'ballpark estimates' of probabilities, is enough to undermine completely the seductive face-value message of the statistical evidence.

There was of course other, medical, evidence in this case, which should not be ignored. The statistical approach to incorporating such additional evidence will be discussed in connection with the Adams case below.

Aftermath

The other statistical expert witness for the defence, Ian Evett of the Forensic Science Service, raised similar issues in his written report for the appeal, including an exposition of the prosecutor's fallacy. However, when Defence Counsel asked leave to call his two statistical experts to testify orally, Lord Justice Henry replied 'We don't need to hear them – it would only be argumentative. After all, it's hardly rocket science.' (Actually, it is. Rocket guidance systems such as those of the Apollo space shots are based on statistical control theory.) So we were denied our day in court, and Sally Clark was denied the opportunity to have the serious logical inadequacies of the prosecution's statistical evidence properly exposed. It was clear from the appeal judgment eventually handed down that much of our written evidence had received only the most cursory attention. Their Lordships claimed that the error in the prosecutor's fallacy was so obvious that it did not even need to be drawn to their attention, and that it could not possibly have had any influence on the trial jury's verdict. The conclusion of the court was that 'any error in the way in which statistical evidence was treated at trial was of minimal significance'. The appeal was dismissed.

Sally Clark was eventually allowed a second appeal, but in this the statistical issues were firmly relegated to the background. It revolved around newly discovered medical evidence of possible bacterial infection in Harry, previously observed but undisclosed by the pathologist, Dr Williams. Although the court did now accept in passing that the original statistical evidence had been misleading, this was done without any new argument, and with frankly questionable logic. At any rate, Sally's conviction was declared unsafe, her second appeal allowed, and her conviction finally quashed on 29 January 2003.

Soon after this, two similar cases hit the headlines. In each, a mother was accused of murdering her babies, the principal prosecution evidence, again proffered by Sir Roy Meadow, being the statistical improbability of such deaths occurring by natural causes. Trupti Patel was acquitted at trial on 11 June 2003; Angela Cannings, convicted in April 2002, was freed on appeal on 10 December 2003. In both cases, the defence brought evidence of similar unexplained infant deaths among the mother's extended family, aiming to establish a possibility that her own children's deaths could have occurred as a result of some genetically heritable trait.

In all three cases there was specific medical or genetic evidence to provide a plausible alternative explanation for the deaths. It is not clear whether, without this, an attack on the statistical evidence alone would have been enough to outweigh its initial impact. But certainly in their aftermath the pendulum has swung to the other extreme. Statistical evidence of the kind presented by Meadow has been thoroughly rubbished in the courts and the press – but again with little evidence of any logical understanding of the real issues – and Meadow is being publicly and professionally hounded for allegedly perverting the course of justice. An urgent review is currently under way of all 298 cases from the past ten years in which a parent or carer was found guilty of murdering a child, as well as of thousands of family law cases where a mother suspected of harming her child has had a child taken into care. At the time of writing, appeals or referrals to the Criminal Cases Review Commission have been recommended for 5 of the 97 completed reviews of criminal cases.

It now seems unlikely that any prosecutions for causing infant deaths, on the basis of 'naked statistical evidence', will be brought to law for the foreseeable future. While this may well be commendable for all sorts of legal and procedural reasons, it would be a great pity if a consequence were to be the banishment of all statistical argument. For example, the impact of medical or genetic evidence could be usefully measured by an estimate of its effect on the all-important ratio between the probabilities of the deaths under natural and unnatural causes.

Identification evidence

One of the major current areas where Statistics impinges on the Law involves the interpretation of identification evidence. In such a case in criminal law, the principal uncertainty is not whether a crime has been committed (the issue for Sally Clark), but rather whether the accused is the culprit. Closely related issues arise in civil law, for example in cases of disputed paternity.

Evidence would now be brought specifically to address the issue of identity. Although this might be, for example, eyewitness evidence, the most incisive identification evidence comes from forensic examination of the crime scene, resulting in the discovery of 'trace evidence' – for example, footprints, fingerprints, cartridge cases, DNA – and its matching to similar information obtained from a suspect person or object. With modern advances in DNA technology, DNA profiling has become by far the most common form of forensic identification evidence brought, and I shall focus on it here. Many (though by no means all) of the logical issues arising from DNA identification are essentially the same as for other kinds of identification evidence. An authoritative account of DNA profiling technology and its forensic applications has been given by its inventor, Sir Alec Jeffreys, in his Darwin College Lecture of 2003.

Initially, a match based on DNA profiling was treated as essentially incontrovertible proof of identity. After all, everyone knows that no two individuals (with the exception of identical siblings or clones) have exactly the same genome, whereas two different samples from the same individual will be genetically identical. However, this ignores serious limitations of both the technology and its naïve interpretation. The forensic scientist could never have the immense scientific, financial and time resources required to construct a full genetic map from a DNA sample, and instead is content with making measurements on a limited number of markers (these being identifiable segments of 'junk DNA' that display significant variation from person to person). But at this more limited level it is no longer impossible for two distinct individuals to share the same DNA profile. This means that a DNA match is no longer complete proof of identity, so that the principal question now becomes: 'What is the strength of the DNA evidence concerning identity?' More generally, how are we to take proper account of the DNA (or similar) identification evidence in the overall context of the case at hand?

One important special feature of DNA identification evidence is the availability of large databases from which the frequency, in the population at large, of the observed measurement on each marker can be estimated with some precision. On top of this, accepted genetic theory justifies us in multiplying such frequencies, across all the measured markers, to calculate the overall 'match probability' – the frequency with which that profile can be expected in the population at large. This step will often result in incredibly small values: figures such as 1 in a billion are now routine.

At this point it could be argued that we are back to the initial state of play. Such incredible rarity appears tantamount to impossibility; so that, when we see a match between DNA profiles obtained from two sources, the only remaining viable explanation must be that they originated from the same individual. But once again, such seemingly obvious arguments from tiny probabilities can be grossly misleading.

I shall illustrate some of the issues involved by reference to specific cases.

Denis Adams

In January 1995 Denis John Adams, who lived in the area of the crime, was tried on a charge of sexual assault. The prosecution case rested on expert evidence of a DNA match between Adams and a sample of semen, extracted from the victim and accepted as being that of the culprit. No other incriminating evidence was presented. The defence relied on the fact that the victim did not identify Adams at an identification parade and said that he did not look like the man who had raped her. In addition Adams' girlfriend testified that he had been with her at the time of the crime.

The prosecution's forensic expert testified that the match probability attached to the DNA evidence was 1 in 200 million. The defence tried to argue that a figure of 1 in 2 million could not be ruled out. It might be thought that, once we enter the realms of such tiny numbers, arguing about just how tiny they are is the equivalent of counting angels on pinheads. But, as we shall see, such argument is not entirely pointless. For the moment, for the sake of argument, we work with the figure of 1 in 2 million.

The logically incorrect but dangerously plausible 'prosecutor's fallacy' is particularly tempting in cases involving identification evidence. It would consist here in misinterpreting the match probability – in fact, *the probability of obtaining a DNA match to the crime sample*, had the culprit been someone other than Adams – as *the probability*, on the basis of the DNA match evidence, *that the culprit was not Adams*. (In fact this argument, though common enough in other cases, was carefully avoided by the prosecution in the case of Adams. We cannot, however, rule out the possibility that the jury nonetheless misunderstood the meaning and impact of the match probability.) More accurately but less memorably, replacing one of these logically quite distinct concepts by the other is also called 'transposing the conditional'.

The first of the two probabilities above refers to the biological and physical processes generating the two DNA profiles found; the second relates specifically to Adams' culpability. Confusion between these quantities is particularly hard to avoid, given that each can be expressed as 'the probability that the crime sample came from someone other than Adams' – a form of words that sounds as if it means something definite but in fact is highly ambiguous. Indeed, natural language sometimes seems to have been carefully crafted to facilitate just this confusion. Robert Matthews, one of the few journalists who understand these issues clearly and take great care in choosing the right words to express them, tells me that his articles are subjected after submission to 'minor editing to improve readability', usually perverting their meaning.

As an analogy to help clarify and escape this common and seductive confusion, consider the difference between 'the probability of having spots, if you have measles' – which is close to 1 – and 'the probability of having measles, if you have spots' – which, in the light of the many alternative possible explanations for spots, is much smaller. Application of the prosecutor's fallacious argument would here have yielded the conclusion that, given the DNA evidence, the probability that Adams is guilty is 1 minus (1 in 2 million) – a number so incredibly close to 1 that there could be no reasonable doubt as to his guilt.

Likelihood ratio

So how should we go about making sense of the DNA evidence?

The task before the jury is to compare two different hypotheses: on the one hand, the *Prosecution Hypothesis*, that the perpetrator of the crime was in fact Adams; and, on the other, the *Defence Hypothesis*, that the perpetrator was someone else. For the sake of argument, we shall assume that under the defence hypothesis we can regard the unknown perpetrator as a random member of some relevant population. (It will often be appropriate to modify or refine this defence hypothesis: for example there may be specific alternative suspects, or one might want to consider the possibility that a – possibly unidentified – relative of the accused was the true culprit. Although such refinements complicate the analysis, they do not affect its overall logic.)

Now, statistical theory has given a great deal of careful attention to the general problem of comparing two hypotheses on the basis of evidence obtained. Although there are a number of schools of thought, with different starting-points, arguments and emphases, all are in agreement that the impact of the evidence can be isolated in a quantity called the *Likelihood Ratio*. This is defined as the ratio of the two probabilities assigned to the given evidence, calculated under the two rival hypotheses. Note that each term in this ratio can be regarded as measuring how well the associated hypothesis explained the data actually obtained: there is at least a superficial resemblance to the philosophical doctrine of 'inference to the best explanation' expounded by Peter Lipton in Chapter 1.

In our context the likelihood ratio arising from the DNA evidence can be expressed as:

$$\frac{\textit{the probability of obtaining the DNA match, if Adams is guilty}}{\textit{the probability of obtaining the DNA match, if Adams is not guilty}}.$$

This is a number that measures the strength of the DNA evidence in favour of the hypothesis of guilt, as against that of innocence. Larger values of the likelihood ratio constitute stronger evidence in favour of guilt. One might consider that the value unity is entirely neutral between the hypotheses, that larger values favour guilt, and that smaller values favour innocence. Although there is a sense in which this is correct, it is a subtle one: as will be discussed below, one must be wary of over-simplistic direct interpretation of the numerical value of the likelihood ratio, which can only be sensibly considered in conjunction with other information.

For interpreting and calculating this expression, a number of background suppositions and items of evidence are typically required. We may suppose that DNA has been taken from the victim that can be assumed to originate from the culprit, and that its profile has been measured and taken into evidence. We also assume that, under the defence hypothesis, the culprit is unrelated to Adams.

The top line of the likelihood ratio is unity: under the assumptions made, and applying current genetic understandings, if Adams is guilty then there has to be a DNA match. As for the bottom line, under the assumptions made, this is just the match probability, which we are taking as 1 in 2 million. Thus the likelihood ratio evaluates to *2 million*.

It seems natural to interpret this very large number as very strong evidence in favour of guilt – an argument distinct from that of the discredited 'prosecutor's fallacy', but tending to much the same conclusion. However, we shall see that matters are not so simple.

Other evidence

Quite apart from its logically misleading nature, a serious problem with the prosecutor's fallacy is that it cannot take account of other, non-DNA, evidence in the case. Whether from a logical, a legal or a common-sense point of view, this is clearly unsatisfactory. In the Adams case all the other evidence pointed towards Adams' innocence, and to ignore it would have been highly prejudicial.

Even in cases where no other evidence is explicitly presented – and this is now common in DNA identification cases – this very fact is significant and should be taken into account. Before introducing any evidence, the accused should be considered as no more likely to be guilty than any other 'random' member of the appropriate population. This might be taken as a mathematical translation of the legal 'presumption of innocence'.

The Adams case was unusual in that both sides took these problems seriously, and attempted to instruct the jury on the reasoning processes needed to address them. In the course of this, numerical values were suggested for probabilities that were not amenable to strict scientific quantification, on the understanding that these values were illustrative, and that each juror should replace them by his or her own assessments before applying the logic. Although this introduces an irreducibly subjective element, there should be reasonable agreement on the order of magnitude of the inputs and the corresponding outputs.

Bayes' theorem

Before any explicit evidence is presented, it might be reasonable to suppose that the culprit is a male aged between about 18 and 60 who is likely to live locally. There were about 150 000 of these, and we could expand this to, say, 200 000 to allow some possibility of a non-local culprit. All that is known about Adams at this point is that he matches these characteristics. Thus the prior probability of his guilt is about 1 in 200 000.

We now face the task of combining this prior assessment with the DNA evidence (we shall consider below the further incorporation of the defence evidence). Fortunately, probability theory tells us exactly how to do this: we have to apply a general result known as *Bayes' Theorem*. In our context this can be expressed as:

$$posterior\ odds = likelihood\ ratio \times prior\ odds,$$

where *prior* [*posterior*] refers to the uncertainty *before* [*after*] incorporating the evidence whose effect is measured by the likelihood ratio. In particular, this shows the very specific status and relevance of the value of the likelihood ratio: it is to effect the journey from prior to posterior uncertainty, rather than, as might be thought, to describe the final destination – which must also depend on the starting point, the prior uncertainty.

We have taken the prior probability of Adams' guilt as 1 in 200 000: this is 1 chance for guilt to every 199 999 against, equivalent to prior odds of 1/199 999 on guilt. The DNA likelihood ratio has already been calculated to be 2 million. Substituting these figures into the right-hand side of Bayes' formula, we calculate the posterior odds on guilt as 2 000 000 × 1/199 999 = 10 (to 4 decimal places). That is, 10 chances of guilt to every 1 chance of innocence, or 10 chances out of a total of 11, giving a posterior probability for guilt of 10/11 = 91 per cent. While high, it would be hard to argue that this is proof 'beyond a reasonable doubt'. In any event, there is a striking difference from the corresponding answer, 1 minus (1 in 2 million), resulting from the prosecutor's fallacy – which would be approximately correct when the prior probability of guilt was 50 per cent, rather than the more appropriate 1 in 200 000 used here.

And now for the defence evidence. This has two separate components:

  i.   the inability of the victim to identify Adams as her assailant;
  ii.   the alibi provided by Adams' girlfriend.

The defence's statistical expert, Peter Donnelly, explained how a juror might go about evaluating likelihood ratios based on these items of evidence, using indicative figures for clarity but pointing out that these should be replaced by the juror's own assessments.

The probability of obtaining the non-identification evidence (i) would be low – say around 10 per cent – if Adams were truly guilty, and higher – say around 90 per cent – if he were innocent. Taking the ratio of these two values produces a likelihood ratio of 1/9 in favour of guilt (which, being smaller than 1, is in fact evidence against guilt).

As for the alibi evidence (ii), we might reasonably expect the girlfriend to produce this, say with probability 25 per cent, even if Adams were guilty; but it would be still more probable, say 50 per cent, if he were truly innocent. This yields a likelihood ratio of 1/2 (recall that it does not matter whether the specific probability figures we have assessed are correct – only their ratio is of importance). Under assumptions that are reasonable in this context, we can multiply together the above component likelihood ratios. The overall defence evidence likelihood ratio becomes 1/18.

Finally, we again use Bayes' formula to update our earlier uncertainty in the light of this further evidence. Multiplying the previously calculated odds of 10 (which, though posterior to the DNA evidence, are prior to the new defence evidence) by the defence's likelihood ratio of 1/18, we find the final odds on guilt, now posterior to all the evidence in the case, to be 5/9: that is, 5 chances for guilt to 9 chances against, or 5 in a total of 14, for a final guilt probability of 5/14 = 35 per cent. If there was cause for reasonable doubt before the defence evidence, after it there can be absolutely no case for conviction.

So far, we have taken the match probability to be 1 in 2 million. Table 6.1 shows the effect of varying this while keeping all other factors unchanged. We see that argument about the number of noughts in the match probability cannot be dismissed as nitpicking. If it can be shown to be as small as 1 in 200 million, the resulting posterior probability becomes 98 per cent, which might be regarded as beyond reasonable doubt, even after having factored in the low

Table 6.1. *The Adams case: dependence of posterior probability of guilt on match probability*

	1 in:		
Match probability:	200 million	20 million	2 million
Posterior probability of guilt:	98%	85%	35%

prior probability and the defence evidence. But the larger values of the match probability are much less convincing.

The above Bayesian argument was presented and accepted without objection at trial, but may have well been lost on the jury, who – perhaps subconsciously swayed by a 'prosecutor's fallacy' interpretation of the match probability – convicted. On appeal a retrial was granted on the basis that the judge had not properly instructed the jury on what to do if it did not wish to follow the Bayesian argument. At retrial the defence again presented the argument, this time against prosecution objection, the jury again convicted, and Adams again appealed. This second time the appeal was dismissed – and with it the whole Bayesian argument, on the grounds that it 'usurps the function of jury' which 'must apply its common sense'. Noble sentiments perhaps; but problematic when common sense can be such a poor guide to handling statistical evidence. Although not a legally binding precedent, this judgment has undoubtedly hampered the presentation of rational statistical argument in the courts.

Hanratty

New issues of calculation and interpretation arise when for some reason it is impossible to obtain a DNA from a suspect. Recourse might then be made to profiling his or her relatives. Because DNA from related individuals will share some features – in a random but well-understood way described by Mendel's laws of genetic inheritance – such indirect profiling is clearly of some relevance: but how, exactly?

In the infamous 'A6 murder' case, James Hanratty was found guilty of murder and rape and hanged on 4 April 1962, going to his death strongly protesting his innocence. From the beginning the verdict was strongly contested, the ensuing disquiet being instrumental in bringing about, in 1965, the abolition of the death penalty in the UK.

Certain items of evidence from the original trial – in particular, a handkerchief found wrapped around the murder gun and knickers from the rape victim Valerie Storie – had been retained by the police ever since. In 1998 it was decided to apply modern DNA profiling technology to re-examine these items. A DNA profile, taken to be from the culprit, was found on both items. Its frequency in the population at large was calculated at around 1 in 2.5 million.

Even though Hanratty was dead and buried and so could not supply a DNA profile for comparison, the popular press greeted the news of this finding in terms such as the following:

> There is a 1 in 2.5 million chance that Hanratty was not the A6 killer.
>
> The DNA is 2.5 million times more likely to belong to Hanratty than anyone else.

The first quotation here has clearly fallen prey to the prosecutor's fallacy. The second might charitably be interpreted as a description of the likelihood ratio, but is more likely to be have been meant, and certainly interpreted, as the posterior odds. But, far more crucially, both have completely missed the point that – with no DNA available from Hanratty – the new DNA evidence from the crime exhibits could be no more incriminating against him than against anyone else!

In an attempt to prove his innocence, Hanratty's mother and brother now offered their own DNA for profiling. Although a full match in these circumstances was not to be expected even had Hanratty been guilty, if there had been some marker at which the crime profile did not overlap at all with that of Hanratty's mother that would have proved that the crime DNA could not have come from him.

In the event there was no such exclusion. This was widely regarded as tantamount to a full match with Hanratty, so justifying use – and misuse! – of the match probability figure of 1 in 2.5 million. Indeed, the forensic expert report did essentially this, referring to a hypothetical suspect whose DNA provided a full match, and thereby erected a prejudicial smokescreen in front of the inconvenient fact that this was simply not the case for Hanratty.

However, while this indirect evidence based on his relatives' DNA clearly points towards rather than away from Hanratty's guilt, measuring its strength is by no means routine or logically straightforward. At one point the forensic report mentions a likelihood ratio figure of 440 (based on the 'STR' component of the DNA). Although no details of this calculation are available, this figure does appear more plausible than the 2.5 million that would be appropriate for a full match. While still providing evidence indicating towards Hanratty's guilt, it is very much weaker than it was widely, and incorrectly, taken to be – especially when we remember that a likelihood ratio is not by itself a measure of certainty in the light of the evidence, but has to be combined with suitable prior odds.

Most of these subtle considerations became redundant when in March 2001 Hanratty's body was exhumed, and it was found that his DNA did indeed provide a full match to the crime profile. With this new evidence the likelihood ratio does now become 2.5 million. Although the defence attempted to attribute the match to contamination during the many years for which the crime items had been stored, they failed to persuade the court, and it would seem that the case is finally closed.

Disputed paternity

Problems of DNA testing for disputed paternity can be regarded as involving indirect matching: of the putative father with the (necessarily unavailable) true father. Given DNA profiles from the mother, child and putative father, the likelihood ratio in favour of paternity can be obtained by calculations well known to forensic geneticists (although these are frequently misinterpreted as supplying the posterior odds). As with all cases of indirect matching, the values so obtained are nothing like the stellar figures typically associated with a direct DNA match, although when based on many markers they can still constitute strong evidence.

A still greater degree of indirectness occurs when the putative father's profile is itself unavailable – perhaps he has fled the country. In that case indirect information about his DNA might be obtained from profiles taken from one or more of his relatives. For example, in one real case DNA profiles were obtained from two full brothers and an undisputed child of the missing putative father, as well as from that child's (different) mother.

Without the principles of probability theory for guidance, forensic scientists have been very unclear as to how to interpret such evidence. With that guidance, attention again properly focuses on the likelihood ratio: we have to compare the

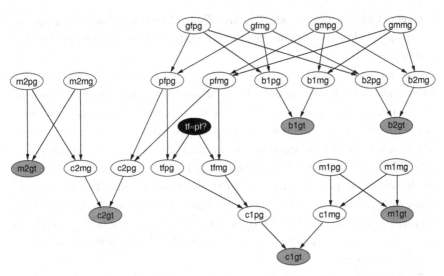

FIGURE 6.1. Probabilistic expert system representation of a complex paternity case.

two probabilities attached to the totality of the observed evidence – whatever its nature – under the competing hypotheses of paternity and non-paternity. But, although this may clarify the logic, calculation of the required probabilities can still be daunting.

Probabilistic expert systems

The modern technology of 'probabilistic expert systems' (PES – also known as 'Bayesian networks') has proved invaluable for solving such problems. A PES is a computer software system that allows one to build a graphical representation of a problem, describe the probabilistic relationships between the variables involved, enter evidence on some of them, and 'propagate' this to obtain revised probabilities for other variables.

Figure 6.1 shows a PES network for the indirect disputed paternity case described above. There is a similar network for each DNA marker measured. The white nodes represent unobserved individual genes, the grey nodes observed genotypes, and the black node the query 'Is the putative father the true father or not?' The arrows indicate probabilistic dependencies, specified numerically elsewhere in the system. On entering and propagating the observed evidence at a given marker, the likelihood ratio this generates can easily be extracted from the query node. The overall likelihood ratio, based on all markers, is obtained by multiplying together all such contributions from individual markers.

For the specific case at hand, 12 genetic markers were used, the resulting single-marker likelihood ratios in favour of paternity ranging from 0.25 to 6.04. While any single one of these is only weak evidence of paternity (and some, being less than 1, even point in the opposite direction), on combining them we obtain a likelihood ratio value of 1303; that is, the overall DNA evidence (on 12 markers for 6 measured individuals) was 1303 times more probable under the hypothesis of paternity than under that of non-paternity. The final step of converting this to a posterior odds or probability cannot be taken without the further input of a prior probability, based on other evidence in the case. If this were, say, 5 per cent, the resulting posterior probability of paternity would be nearly 99 per cent.

There are many other variations on the basic theme of DNA profiling, where both the logical and the computational difficulties of interpretation are magnified still further. These include issues such as: multiple perpetrators and/or stains; mixed crime stains (as in rape, or scuffle); database search to identify a suspect; mutation; contamination; laboratory errors, etc. Some of the purely computational problems arising can again be handled using PES. As an example, Figure 6.2 shows a PES network that can account for the possible disturbing effect of genetic mutation on attributions of paternity, as well as supplying estimates of mutation rates based on nuclear family data when the possibility of non-paternity has to be allowed for. Here white and grey nodes are much as before, the query node is black, horizontally striped nodes model the mutation process, vertically striped nodes the overall possibly unknown mutation rate, and diagonally striped nodes various relationships of genetic compatibility or incompatibility among the family profiles.

Mixed evidence

Graphical representations, such as a PES, can also be invaluable at a purely qualitative level, in helping us to organize and comprehend complex webs of relationships between multiple items of evidence from a variety of sources. The following fictitious but realistic example from a paper by I. W. Evett and myself combines eyewitness, fibre and blood evidence:

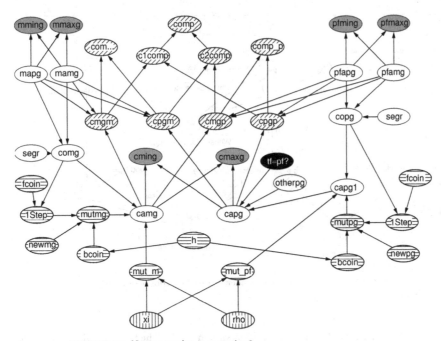

FIGURE 6.2. Non-paternity or mutation?

An unknown number of offenders entered commercial premises late at night through a hole, which they cut in a metal grille. Inside, they were confronted by a security guard who was able to set off an alarm before one of the intruders punched him in the face, causing his nose to bleed.

The intruders left from the front of the building just as a police patrol car was arriving and they dispersed on foot, their getaway car having made off at the first sound of the alarm. The security guard said that there were four men but the light was too poor for him to describe them and he was confused because of the blow he had received. The police in the patrol car saw the offenders only from a considerable distance away. They searched the surrounding area and, about 10 minutes later, one of them found the suspect trying to 'hot wire' a car in an alley about a quarter of a mile from the incident.

At the scene, a tuft of red fibres was found on the jagged end of one of the cut edges of the grille. Blood samples were taken from the guard and the suspect. The suspect denied having anything to do with the offence. He was wearing a jumper and jeans, which were taken for examination.

A spray pattern of blood was found on the front and right sleeve of the suspect's jumper. The blood type was different from that of the suspect, but the same as that from the security guard. The tuft from the scene was found to be

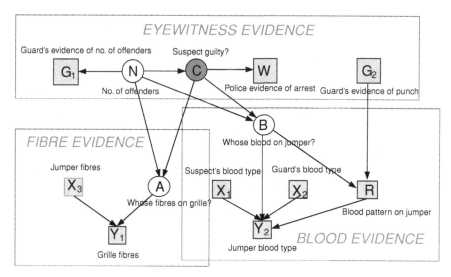

FIGURE 6.3. A case of mixed evidence: PES for robbery example.

red acrylic. The suspect's jumper was red acrylic. The tuft was indistinguishable from the fibres of the jumper by eye, microspectrofluorimetry and thin layer chromatography. The jumper was well worn and had several holes, though none could clearly be said to be a possible origin for the tuft.

Figure 6.3 captures the salient features of this story and their probabilistic causal relationships. As in the DNA networks above, it has been helpful to introduce some unobserved nodes, here N, A and B, as well as the 'query node' C. The network could be elaborated by numerical specification of the probabilistic dependencies indicated by the arrows. But even without this it is possible to read off from the network, by well-defined rules, certain purely qualitative relationships. For example, it can be deduced that once we know N (the number of offenders) and A (who left the fibres on the grille), any further information carried by G_1 (what the guard said about N) and Y_1 (the properties of the fibre tuft) is entirely independent of that carried by R (the pattern of blood on the jumper), and totally uninformative about B (the identity of the person who punched the guard). Such relationships between items of evidence not only are of interest in themselves, but also can be used to simplify the expression and calculation of relevant likelihood ratios.

Even in more straightforward cases, a full and careful analysis will typically require considerable elaboration of the hypotheses and the evidence and of the relationships between them. Thus, evidence of Sally Clark's opportunity to commit murder is not of itself evidence of exclusive opportunity (in particular, Stephen Clark had also been suspected of causing at least one of the deaths), still less of her criminal intent to murder – a necessary legal component of her guilt. This could itself be separated into separate charges over each death; while the alternative hypothesis of SIDS was never really anything other than a 'straw man', and (as transpired at second appeal) by no means exhausts all possible medical explanations. Similarly, DNA evidence of sexual contact can never in itself be evidence of rape, still less, in Hanratty's case, of having murdered someone else; more generally, identification evidence which places an individual at a crime scene does not in itself entail criminal behaviour or guilt. Such hierarchies of propositions and their relationships with the evidence can be handled algebraically in simple cases, but any degree of complexity is more helpfully represented and manipulated using graphical aids.

Wigmore chart

The idea of constructing graphical representations to structure and simplify complex problems has a long history in Law. A seminal contribution was the 'chart method' introduced by the American jurist John Henry Wigmore in 1913 as an aid to trial lawyers in preparing for and acting in court. Although ignored for many years, its usefulness and generality are now better appreciated, thanks in particular to the efforts of William Twining, Terence Anderson and David Schum.

In this approach the problem is first broken down into known and unknown atomic propositions, specified in a 'key list'. These are then represented as nodes of a graphical network, and joined by arrows describing various qualitative forms and strengths of evidential relations between them. Wigmore used a wide variety of shapes and other annotations to describe the various forms of evidential relationship: an example from his book is reproduced as Figure 6.4. Modern versions operate with a reduced collection.

There are both similarities and differences between the PES and Wigmorean charting methods. As for similarities, the Wigmorean approach explicitly aims

§ 33. **Same: an Example Charted.** We shall thus have charted the results of our reasoning upon the evidence affecting any single probandum. But this probandum will usually now in its turn (*ante*, § 8) become an evidentiary fact, towards another probandum in a catenate inference. The process of charting and valuation has then to be renewed for this new probandum; and so on until all the evidence has been charted, and the ultimate probanda in issue under the pleadings have been reached.

The following portion of a chart will illustrate (taken from the case of *Com.* v. *Umilian, post,* § 38):

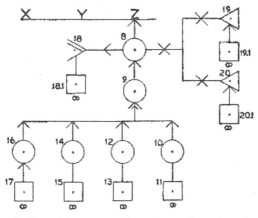

Z is one of the ultimate probanda under the pleadings, viz. that the accused killed the deceased. Circle 8 is one of the evidentiary facts, viz., a revengeful murderous emotion. The arrowhead on the line from 8 to Z signifies provisional force given to the inference.

FIGURE 6.4. An original Wigmore chart (from Wigmore, 1937).

to represent a specific individual's standpoint rather than 'objective truth'; typically a PES can helpfully be considered as serving the same purpose. Both approaches organize many disparate items of evidence and their relationships, focus attention on required inputs, and support coherent narrative and argumentation.

A PES works with variables and questions, distinguishes between causal and inferential relationships, answers relevance queries, supports efficient computation and simplifies expression of likelihood ratios and posterior odds. A Wigmore chart works with propositions, focuses on inference towards some ultimate probandum, emphasizes the distinction between occurrence and report of an event, pays particular attention to the many links in a chain of reasoning, and assists qualitative analysis and synthesis. Current work is exploring these similarities and difference with the aim of developing a system combining the best features of both approaches.

As a comparative case study, David Schum has constructed a Wigmore chart for the fictitious robbery case previously modelled by a PES in Figure 6.3. Working from the standpoint of the prosecutor, he takes as the ultimate probandum:

U: Harold S. unlawfully and intentionally assaulted and injured a security guard Willard R. during a break-in at the Blackbread Brewery premises, 27 Orchardson St., London NW8 in the early morning hours of 1 May 2003.

This is dissected into penultimate probanda:

P₁: In the early morning hours of 1 May 2003, four men unlawfully broke into the premises of the Blackbread Brewery, located at 27 Orchardson St., London NW8.

P₂: Harold S. was one of the four men who broke into the premises of the Blackbread Brewery in the early morning hours of 1 May 2003.

P₃: A security guard at the Blackbread Brewery, Willard R., was assaulted and injured during the break-in at the Blackbread Brewery on 1 May 2003.

P₄: It was Harold S. who intentionally assaulted and injured Willard R. during the break-in at Blackbread Brewery on 1 May 2003.

Schum's full key list contains 97 propositions, which number could have been further expanded by explicit incorporation of the generalizations used to warrant evidential links between other items. Table 6.2 contains the subset of these propositions relating to penultimate probandum **P₂**: this subset is charted (using leaner modern symbolism) in Figure 6.5.

A general approach to evidence

I have discussed statistical reasoning, probabilistic expert systems and Wigmore charts in the specific context of Law, but in fact these are completely general

Table 6.2. *Robbery. Key list for penultimate probandum P$_2$*

(29)	The intruders' car left immediately at the first sound of the alarm, leaving the intruders stranded.
(30)	Willard R. testimony to (29).
(31)	The intruders dispersed from the Blackbread Brewery premises on foot.
(32)	Willard R. testimony to (31).
(33)	The four intruders went their separate ways.
(34)	In a search of the area surrounding the Blackbread Brewery premises, police apprehended Harold S. trying to 'hot wire' a car in an alley about $\frac{1}{4}$ mile from the Blackbread Brewery premises.
(35)	DI Leary testimony to (34).
(36)	A photo of Harold S. taken shortly after his apprehension to be shown at trial.
(37)	The photo shown at trial is the same one police took of Harold R. shortly after his arrest.
(38)	The car Harold S. was trying to 'hot wire' did not belong to him.
(39)	Harold S. was one of the four intruders fleeing the Blackbread Brewery premises.
(40)	During the police investigation a short time after the intrusion, the police found a tuft of red fibres on a jagged end of one of the cut edges of the metal grille on the Blackbread premises.
(41)	DI Leary testimony to (40).
(42)	The tuft of fibres to be shown at trial.
(43)	The tuft of fibres shown at trial is the same one that police found on a jagged end of one of the cut edges of the metal grille on the Blackbread premises.
(44)	The tuft of the fibres found on the metal grille on the Blackbread Brewery premises is red acrylic.
(45)	DI Leary testimony to (44).
(46)	The tuft of red acrylic fibres found on the metal grille came from an article of clothing.
(47)	The article of clothing the fibres came from was being worn at the time of the break-in at the Blackbread Brewery.
(48)	Harold S. was wearing a jumper and jeans at the time of his apprehension.
(49)	DI Leary testimony to (48).
(50)	The jumper and jeans to be shown at trial.
(51)	The jumper and jeans to be shown at trial are the same ones the police took from Harold S. after his apprehension.
(52)	Harold S.'s jumper is made of red acrylic.
(53)	DI Leary testimony to (52).
(54)	Harold S. was wearing this red acrylic jumper at the time of the break-in at Blackbread Brewery.

Table 6.2. *(cont.)*

(55)	The tuft of red fibres found on the metal grille on the Blackbread Brewery premises is visually indistinguishable from the fibres on Harold S.'s jumper.
(56)	DI Leary testimony to (55).
(57)	The tuft of fibres and the jumper to be shown together at trial.
(58)	The tuft of fibres and the jumper are the same ones police obtained during their investigation of the break-in at the Blackbread Brewery.
(59)	The tuft of red fibres found on the metal grille on the Blackbread Brewery premises is indistinguishable from the fibres on Harold S.'s jumper as shown by a microspectrofluorimetry analysis.
(60)	DI Leary testimony.
(61)	Microspectrofluorimetry analysis result to be shown at trial.
(62)	The microspectrofluorimetry results shown at trial are the same ones police obtained from the forensic scientist ['boffin'] who performed the analysis.
(63)	The tuft of red fibres found on the metal grille on the Blackbread Brewery premises is indistinguishable from the fibres on Harold S.'s jumper as shown by a thin layer chromatography analysis.
(64)	DI Leary testimony to (63).
(65)	The results of the thin layer chromatography analysis to be shown at trial.
(66)	The thin layer chromatography results shown at trial are the same ones police obtained from the forensic scientist who performed the analysis.
(67)	The jumper belonging to Harold S. is well worn and has several holes in it.
(68)	DI Leary testimony to (67).
(69)	None of holes in Harold S's jumper can be clearly identified as a possible source of the tuft found on the metal grille on the Blackbread Premises.
(70)	DI Leary testimony to (69).
(71)	Matching of tufts to holes in fabrics is very difficult.
(72)	The jumper worn by Harold S. on 1 May 2003 was torn on a hole in the metal grille at the Blackbread premises.
(73)	Harold S. was wearing the article of clothing that produced the tuft of red acrylic found on a jagged end of the hole cut into the metal grille at the Blackbread Brewery premises on 1 May 2003.
(74)	Testimonial denial by Harold S. of P2, that he was one of four men who broke into the premises of the Blackbread Brewery in the early morning hours of 1 May 2003.

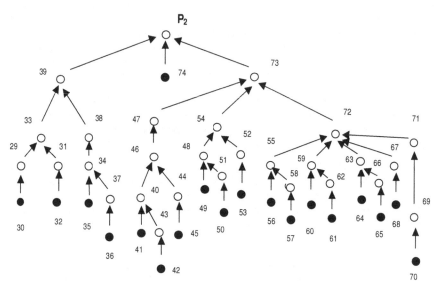

FIGURE 6.5. Wigmore chart for robbery.

formal tools for representing, manipulating and interpreting evidential relationships, with a vast array of potential applications. There is no reason why
they could not be fruitfully applied much more widely. As one example of this
broader viewpoint, Figure 6.6 shows part of a Wigmore chart constructed by
Terence Anderson to address a historical query raised by Mark Geller: When
did the ability to read cuneiform script disappear?

Still more broadly, such extensions suggest that it should be valuable to try
and identify general logical principles underlying the interpretation of evidence
across all fields of human enquiry, together with general tools for applying
them. It is remarkable that, while the need for such an approach to evidential
analysis is at least as old as the need for Aristotelian logic, and arguably even
more pressing, neither the ancient Greeks nor their modern counterparts have
seen fit to pay it the same degree of attention.

The perception of this need has provided the impetus for an interdisciplinary research programme, 'Evidence, Inference and Enquiry', which has
recently been established at University College London. This is involving participants with a wide range of disciplinary backgrounds and affiliations, including
Statistics, Law, Crime Science, Psychology, Economics, Philosophy of Science,

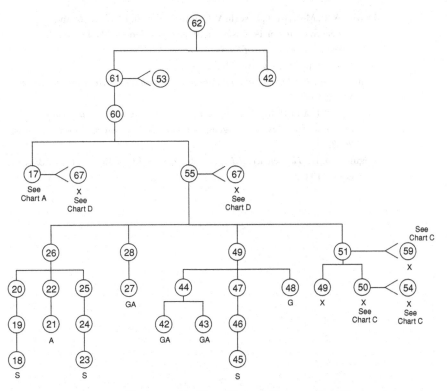

FIGURE 6.6. 'The last wedge: When did the ability to read cuneiform script disappear?' Wigmore chart B, addressing penultimate probandum (62): *Iamblichus knew and could read Akkadian cuneiform until the beginning of the third century AD.*

Geography, Medicine, Ancient History, Computer Science and Education. Topics being addressed include: subject- and substance-blind approaches to evidence; inference, explanation and causality; recurrent patterns of evidence; narrative, argumentation, analysis and synthesis; cognitive biases; formal rules; decision aids; and interdisciplinary comparisons. There is much ground to be covered, but the journey has begun.

FURTHER READING

Anderson, T. J. and Twining, W. L., *Analysis of Evidence*, London: Weidenfeld and Nicholson, 1991.

Dawid, A. P., 'Bayes's Theorem and Weighing Evidence by Juries', *Proceedings of the British Academy* 113, 2002, 71–90.

Dawid, A. P., Mortera, J., Pascali, V. L. and van Boxel, D. W. , 'Probabilistic Expert Systems for Forensic Inference from Genetic Markers', *Scandinavian Journal of Statistics* 29, 2002, 577–95.

Sally Clark Supporters (www.sallyclark.org.uk/).

Schum, D. A., *The Evidential Foundations of Probabilistic Reasoning*, New York: Wiley, 1994.

Twining, W. L. and Hampsher-Monk, I., *Evidence and Inference in History and Law: Interdisciplinary Dialogues*, Evanston: Northwestern University Press, 2003.

Wigmore, J. H., *The Science of Judicial Proof*, 3rd edition, Boston: Little, Brown, & Co., 1937.

7 Legal Evidence: Judging the Verities of Advocates

CHERIE BOOTH

Legal evidence: standards of truth

In a court of law, the truth may be elusive. Adversarial proceedings pit opposing parties against each other, each of whom presents rival conceptions of reality. In *The Picture of Dorian Gray*, Oscar Wilde noted the value of putting our notions of truth to the test: 'The way of paradoxes is the way of truth. To test Reality we must see it on the tight-rope. When the Verities become acrobats we can judge them.' In the courtroom, advocates use legal evidence to paint a picture of the dispute – capturing their own conception of reality. The court's function is to judge the relative verities of the paradoxical portraits. The portrait that captures reality more completely is the basis upon which legal liability, and the rights of the parties, are determined.

Darwin's Theory of Natural Selection: proof by probabilities

The elusiveness of truth was also known to Charles Darwin. To demonstrate the truth of his theory of natural selection, Darwin appealed to its explanatory value in a letter to the botanist J. S. Henslow in 1860:

> Why may I not invent a hypothesis of natural selection (which from analogy of domestic productions, and from what we know of the struggle of existence and of the variability of organic beings, is in some very slight degree in itself probable) and try whether this hypothesis of natural selection does not explain (as I think it does) a large number of facts in geographical distribution – geological succession – classification – morphology, embryology, etc., etc.
>
> (*Correspondence of Charles Darwin* VIII, ed. F. Burkhardt, 1993, 195).

In accordance with this methodology, Darwin induced the existence of a general state of affairs from a series of acknowledged and particular facts. Where multiple states of affairs exist, each of which may explain the observable facts, the scientific mind utilizes probability to identify the superior validity of the theory that is then held to be true. John Stuart Mill recognized the prima facie veracity of Darwin's theory, and in 1860 told Henry Fawcett, a Cambridge economist, that: 'though he cannot be said to have proved the truth of his doctrine, he does seem to have proved that it *may* be true, which I take to be as great a triumph as knowledge and ingenuity could possibly achieve on such a question'.

However, in *Three Essays on Religion*, Mill finds that Darwin failed to discharge the scientific standard of proof, and concludes that the 'adaptations in Nature afford a large balance of probability in favour of creation by intelligence'.

In his revolutionary treatise, *The Logic of Scientific Discovery*, Karl Popper questioned the validity of the inductive method of proof to which Mill and his Victorian philosopher colleagues subscribed, and against which Darwin's theory was initially critiqued:

> It is far from obvious, from a logical point of view, that we are justified in inferring universal statements from singular ones, no matter how numerous; for any conclusion drawn in this way may always turn out to be false: no matter how many instances of white swans we may have observed, this does not justify the conclusion that *all* swans are white.

Instead, Popper relies on the notion of 'falsifiability' as the 'criterion of demarcation':

> According to my proposal, what characterizes the empirical method is its manner of exposing to falsification, in every conceivable way, the system to be tested. Its aim is not to save the lives of untenable systems but, on the contrary, to select the one which is by comparison the fittest, by exposing them all to the fiercest struggle for survival.

The scientist's understanding of 'reality' therefore depends upon a falsification of competing theories about the world. The methodology of scientific proof, embodied in the theories of Popper, is integral to the workings of the judicial process and the protection of individual rights.

The burden of proof: disproving innocence

The structure of our common law legal system mandates that the freedom of the individual is best secured by a set of laws that interfere with the freedom of an individual only to prohibit conduct that inhibits the freedom of others. The law is prohibitive, and not permissive. Our law does not assume that we are liable for the doing of an act unless we prove that the law permitted us to do that act. Instead, the 'natural state' of an individual's legal liability is innocence. The lawyer 'sees' the legal liability of an individual 'in a different way', so that 'the new fact' of guilt is proved, only when the old fact of innocence is disproved. This right is enshrined in Article 6(2) of the European Convention of Human Rights, which mandates that '[e]veryone charged with a criminal offence shall be presumed innocent until proved guilty according to law'. Similarly, in civil cases, the plaintiff bears the onus of proving that the defendant's acts give rise to civil liability. However, at this point, the fundamental difference between scientific and legal methods of proof becomes apparent: this being the different standards of certainty to which hypotheses must be disproved.

Standards of proof: particular justice

Scientists disprove hypotheses by experiments whose results may be measured statistically. Results from these experiments are deemed to be 'statistically significant' if, but only if, there is a less than 5 per cent chance that the result was obtained by chance alone and not due to an actual association with the subject of study. Courts of law similarly rely upon notions of probability to judge the relative truth of competing arguments. Standards of proof are the measures of certainty at which courts are prepared to impose legal liability. In criminal cases, the courts require significant certainty of the truth. Sir William Blackstone stated that: 'It is better to let ten guilty men go free than to wrongly incarcerate one innocent man.'

The presumption of innocence mandates that 'it is the duty of the prosecution to prove the prisoner's guilt' beyond reasonable doubt. This means (as was argued in *Woolmington* v. *DPP* in 1935) that: 'If, at the end of and on the whole of the case, there is a reasonable doubt, created by the evidence given by either the prosecution or the prisoner, as to whether the prisoner [committed the crime] the prosecution has not made out the case and the prisoner is entitled to an acquittal.' In civil cases, the courts endeavour to maintain a balance between the parties by imposing upon the plaintiff the burden of proving the defendant's liability on 'the balance of probabilities'. In his book, *The Probable and the Provable* (1977), Jonathan Cohen identifies the potential for injustice that is inherent in this standard of proof:

> Suppose the threshold of proof in civil cases were judicially interpreted as being at the level of a mathematical probability of 0.501. Would not judges thereby imply acceptance of a system in which the mathematical probability that the unsuccessful litigant deserved to succeed might sometimes be as high as 0.499? This hardly seems the right spirit in which to administer justice.

However, the cause for concern is somewhat alleviated by the law's disregard of mathematical probability. Cohen highlights that 'Pascalian' probability is an inappropriate tool for the judicial fact-finding exercise. This is because it calculates the probative value of evidence by multiplying the degree of uncertainty with which one piece of evidence is to believed by the degree of uncertainty with which another is to be believed. This calculation is plainly inconsistent with the courts' evidential tool of corroboration, whereby the testimony of one witness to the effect that the accused was present at the scene of the crime is used to support the similar testimony of another witness, even if the vision of the first witness was obstructed whilst the demeanour of the second witness in the courtroom was dubious. The problem then with Pascalian probability is that: 'Every assignment of a Pascalian probability to a particular conclusion rests on an assumption that the premisses contain *all* the relevant facts.' Popper's 'Formal Theory of Probability' overcomes this hurdle. As Popper notes in *The Logic of Scientific Discovery*, the autonomy of his theory mandates that 'probability conclusions can be derived only from probability premises'. Cohen explains that: 'when faced with this apparent plurality of criteria, [the theory] seems content to pursue the ideal of step-by-step descriptive adequacy rather than that of explanatory simplicity'. Yet, as courts are obliged to reach a verdict, regardless of the incompleteness of the evidence admitted by the parties, the formality and autonomy of the theory make it a tool of limited utility for courts of law. Consequently, courts rely upon a method of reasoning that differs from our common conceptions of probability, utilized in the tossing of a coin, as well as from the sophisticated structures of proof utilized in disproving scientific theories. Standards of legal proof involve assessments of *rational* probability. Statistics are replaced (as Glanville Williams has noted) by 'our experience of past (and present) events'. Expressed at its most analytically intense, Cohen, writing in the 1991 *Criminal Law Review*, describes 'the proper structure of forensic proof' in Anglo-American legal systems as 'Baconian', so that where P on balance supports Q, 'the strength of an inference from P to Q [is graded] . . . by grading the extent to which P does state all the relevant facts – i.e. the extent to which P eliminates lurking reasons for doubt'.

Like Darwin's Theory of Natural Selection or an experiment whose results are statistically significant and tend to disprove a hypothesis, the value of any legal argument lies in the feasibility of its explanation of all the relevant facts, all the admissible evidence. By contrast, the standard of proof that courts require is not the standard demanded by scientists. The focus of scientists is the attainment of a body of general knowledge, whilst the focus of the courts is the doing of particular justice. In civil cases, the law is concerned with justice as between the parties. It is not concerned with general notions of justice, or of ensuring that justice is fairly distributed amongst a general population. In criminal cases, the law is concerned with ascertaining the guilt of the accused, and generally does so by concerning itself with only the circumstances of the alleged crime, and not either the accused's general tendencies or the impact upon society of the court's verdict. This is why not all the evidence is admissible – for example evidence of the accused's criminal propensity or past convictions – and it is also why an 'informational vacuum' may be constructive, and not destructive, of a fair trial.

It is this difference in purpose, the court's exclusive focus upon the circum-stances of the individual parties before it, that sets the legal methodology apart from that employed by scientists. The following legal analysis examines this difference by looking at the ways that legal evidence, and the methods of legal proof, are shaped by the unique purpose of judicial inquiries: that purpose being to determine the criminal responsibility or civil liability of the parties that appear before the court. The analysis concentrates on three topical legal issues.

Firstly, the Domestic Violence, Crime and Victims Bill proposes reforms to integrate the civil and criminal proceedings through which victims of domestic violence seek protection. Despite the need to streamline judicial processes in this field, there is a concern that the vesting of criminal courts with powers to issue civil restraining orders upon an acquittal mixes the civil and criminal burdens of proof within singular proceedings and thereby infringes principles of a fair trial.

Secondly, with reform to the law of corporate manslaughter pending, the analysis considers the role that scientific evidence plays in determining the issue of causation in the increasingly litigious field of toxic tort litigation. Of particular interest is the effect that the difference in scientific and legal standards of proof has upon the relevance of scientific evidence to legal proceedings.

Thirdly, I examine the exceptional role that international criminal tribunals perform as mechanisms for declaring the unacceptability of certain conduct and fostering a culture of accountability. It is conjectured that that role necessitates the tribunals' adoption of methods of proof, the primary purpose of which is not to ascertain the truth with respect to each atrocity, but to establish the legitimacy of the tribunals as international and impartial dispensers of justice.

Legal evidence in practice: civil and criminal standards of proof

One of the aims of the recently introduced Domestic Violence, Crime and Victims Bill is to streamline the myriad of trial processes through which imperilled partners seek protection. Currently, a victim (or potential victim) of domestic violence may institute civil proceedings pursuant to either section 3 of the Prevention from Harassment Act 1997, to obtain an injunction to restrain the 'defendant from pursuing any conduct which amounts to harassment', or Part IV of the Family Law Act 1996 (and, in particular, section 42), to obtain a 'non-molestation order' to prohibit the respondent from molesting the victim, or 'a relevant child'. These applications are resolved by the Family Proceedings Court, circuit and district judges of the County Court and High Court judges of the Family Division. Further, the Director of Public Prosecutions may institute prosecutions against a perpetrator of domestic violence for criminal charges enumerated in the Prevention from Harassment Act 1997 and the Offences against the Person Act 1861. The guilt of the accused is adjudicated upon by the adult Magistrates' Court or the Crown Court. The multiplicity of jurisdictions from which urgent relief may be sought diminishes the capacity of the legal system to provide effective protection to victims of domestic violence. It does so in three ways. First, the relief granted by different courts in different jurisdictions may be inconsistent. For example, the bail conditions concerning residence that a court imposes on an alleged perpetrator may frustrate a non-molestation order. Second, the victim is required to make multiple court appearances. In presenting the measures of the Bill, Harriet Harman QC MP recounted in December 2003 the unfortunately common occurrence where 'The man had been acquitted but the police were so worried about the woman's safety that they had to give her a police escort to the civil court to seek an injunction.' Third, evidence of the same facts is presented to different judges repeatedly, and often – for the victim – traumatically. In the words of Deborah Epstein (writing in the *Yale Journal of Law and Feminism*), each judge resolves the application in 'an informational vacuum'.

The proposed Bill seeks to address these concerns by enabling the criminal courts to grant a civil restraining order against an accused who is acquitted of an offence. The order may be granted if the court 'considers it necessary to do so to protect a person from harassment by the defendant', and the order may prohibit 'the defendant from doing anything described in the order'.

Mixing of standards of proof: causes for concern

This cross-vesting of civil jurisdiction in criminal courts involves the mixing of standards of proof. The difference between the criminal and civil standards makes it possible, and logically sound, for a jury to consider an accused not guilty beyond reasonable doubt, but a judge to consider a defendant an unacceptable threat to the alleged victim. In this way, the criminal proceedings are 'like the use of a measuring-tape marked off in centimetres as well as in inches when these different criteria of length are defined in terms of a standard metre-bar and a standard yard-bar, respectively' (Cohen, 1977). However, the proceedings of a criminal trial are an inappropriate measuring-tape for assessing the appropriateness of the court granting civil remedies. This is because the facts at issue in criminal proceedings are not necessarily at issue in civil proceedings.

Criminal guilt depends upon proof that the accused committed all the elements of the offence. A necessary element of a criminal offence is a criminal act. In defending his innocence, the criminal standard of proof beyond reasonable doubt entitles an accused to conduct an inactive defence, which merely compels the Prosecution to 'come up to proof'. On the other hand, an accused may pursue an active defence, where, for reasons of tactics at trial, only one of the multiple elements of an offence is disputed. The disputed element may, or may not, be that the accused committed the act. The burden of proof beyond reasonable doubt entitles the accused to leave issues uncontested.

On the other hand, the proposed restraining order does not require proof that the alleged perpetrator committed any act, but rather that the defendant is likely, in the future, to subject the applicant to harassment. The emphasis is on the long-term safety and security of the victim, and not the imposition of justice for past wrongs.

The problem intrinsic in making this assessment immediately subsequent to the dismissal of criminal charges is that it is possible that the judge, in making the assessment, may draw the inference that the accused's lack of contest of elements of the criminal offence constitutes an admission of guilt. To an extent, this issue appears to be resolved by the Bill's proposed procedural measure, whereby both the prosecution and the defence are entitled to lead any further evidence upon the jury's verdict of 'not guilty'. The accused is therefore free to lead any evidence that he has withheld from the jury. However, to the extent that the restraining order is granted by a judge in the aftermath of a possibly fierce courtroom battle, in which the accusatory dynamics are (unlike some Family Court proceedings) one-way, there is potential for prejudicial crystal-ball gazing.

That the courts must make an assessment of the likelihood of the defendant's future conduct, and not the likelihood that he had previously committed an act, is evident from the purpose and wording of the provisions of the Bill. In the decision of *R (on the application of McCann and others)*, the House of Lords interpreted the provisions of section 1 of the Crime and Disorder Act 1998, which relate to the making of anti-social behaviour orders, and concluded that: 'The inquiry under s1(1)(b), namely that such an order is necessary to protect persons from further anti-social acts by him, does not involve a standard of proof: it is an exercise of judgment or evaluation.' Even where the accused denies that the incidents that are the subject of the criminal charges occurred, it is possible that the judge's 'judgment or evaluation' may be coloured by the fact that the accused's conduct was sufficient to warrant his being suspected and charged with criminal offences. Darwin's observations of human nature, as recorded in his *Notebooks*, are pertinent here: 'Do not our necessary notions follow as consequences on *habitual* or instinctive assent to propositions, which are the result of our senses, or our experience?'

Whilst our senses, and research in the area (like that of Professsor Epstein from the Washington Domestic Violence Initiatives), tell the sad tale that most (68 per cent) of those charged with domestic violence offences are guilty and a restraining order is, therefore, necessary, the merits of each restraining order ought to be assessed independently. The demands of a fair trial necessitate that our judges reach conclusions of fact in 'informational vacuums', whereby the satisfaction of the relevant standard of proof is adjudged by reference to only the relevant and admissible legal evidence.

An alternative: integration of courts, division of judges

The advantages, especially to victims of domestic violence, of streamlined court procedures are easily observed. In November of 2003, I accepted an invitation to visit the Domestic Violence Unit of the Superior Court in Washington DC. I observed the expediency with which the judge of the Court was able to call for both the civil and criminal files to be before him and to make pre-trial orders which ensured that the proceedings ran smoothly together. Bail conditions mirrored the family courts' child contact orders, while the judge was able to list the proceedings so that the criminal trial began quickly and immediately before the final family hearing. All the advocates in both proceedings were before the Court, making the pre-trial procedure run much more smoothly.

In the Washington courts, as well as those in Florida and Hawaii, the stream-lining of domestic violence court proceedings is even more advanced than the proposals of the Domestic Violence Bill. As Epstein has noted, the creation of 'integrated domestic violence courts' is based upon a policy of 'one family, one judge', so that proceedings concerning criminal charges and civil protection orders are combined with family law.

The President of the Family Division has proposed the creation of similarly integrated courts in England. Specifically, Dame Elizabeth Butler-Sloss suggests that domestic violence courts be presided over by a circuit judge, district judge or bench of magistrates (able to hear both the family and the criminal aspects of domestic violence). The criminal and civil proceedings would be heard sequentially, maintaining the separation of proceedings and rules of evidence. Whilst there may be objections to the same judge hearing both proceedings, in my view any objection made on that basis would not be a sensible one unless there were some other good reason for it.

One reason for not objecting, perhaps, is that the achievements of the Domestic Violence Unit of the Washington Superior Court may be replicated by the creation of an integrated court, within the confines of a single court building, without requiring the services of a single judge. The integrated court may exercise civil, family and criminal jurisdictions so that information is shared between prosecutors and victims, the listings for civil, family and criminal proceedings are synchronized, and the terms of restraining orders, bail conditions and child contact orders are consistent. A proceeding for a restraining order may be determined immediately subsequent to the return of a jury by a judge sitting in the civil jurisdiction, but sitting in the next-door courtroom. Rules of civil procedure may mandate the prosecuting authorities, upon a bail application, to furnish all information concerning pre-existing restraining orders and child contact orders. Protection for victims of domestic violence may be improved, and the rules of procedural fairness for the accused and defendant maintained, by the adoption of a policy similar to, though slightly amended from, that of the District of Columbia: 'one family, one court, many judges'.

So far, I have looked at the importance of the courts maintaining the distinction between the civil and criminal standards of proof, by ensuring that the civil proceedings are resolved by a trier or assessor of fact who is seen as being independent of, and whose perception has not been tainted by, the initial criminal proceedings. The distinction is important because it enables the courts to focus on the merits of each individual proceeding. Within each proceeding, the trier of fact is also required to focus upon the merits of each argument that the parties posit. In cases where scientific evidence is admitted to demonstrate the probable truth of an argument, the distinction between legal and scientific standards of proof ensures that the courts focus upon the reasonableness of the parties' competing hypotheses, and not the value of the hypotheses as scientific theorems.

Legal and scientific standards of proof

Issues of proof surrounding alleged violence in the home arose, in a different form, in the tragic case of *R* v. *Sally Clark*. Ms Clark was convicted for the murder of her two children, who were aged 11 weeks and 8 weeks, respectively. At trial, the prosecution case was that: 'The circumstances of both deaths shared similarities which would make it an affront to common sense to conclude that either death was natural, and it was beyond coincidence for history to so repeat itself.'

The defence contended that both children died of natural causes – namely, Sudden Infant Death Syndrome (SIDS). The prosecution led statistical expert evidence that the probability of two SIDS deaths in one family matching the profile of the appellant were 1 in 73 million. One of the defence's grounds of appeal was that: 'The evidence given by Professor Meadow of the statistical probability of two SIDS deaths in one family undermined the safety of the convictions, in that . . . the judge failed to warn the jury against the "prosecutor's fallacy" in relation to the use of statistical evidence.' On appeal, the Court of Appeal acknowledged that, where the fact at issue is the commission by the accused of an act, as in Sally Clark's case, or proof as to the identity of an assailant, as in the case of DNA evidence, statistical evidence is of limited relevance. This is because of the 'prosecutor's fallacy', whereby lay lawyers and juries fallaciously assume that the statistics disclose the probability that the defendant is innocent. Instead (as Balding and Donnelly have argued), the statistics disclose merely the probability that a family may suffer two deaths from SIDS, or the probability that two DNA samples may match.

The limited relevance of statistical evidence in proving identity or commission of an act is highlighted by the submission of statistician Philip Dawid to the Court of Appeal in Sally Clark's case. Professor Dawid calculated the defence's counter-statistic, and found that the probability of two children from the same family being murdered was 1 in 2 billion. This 'counter-statistic' makes the death of a second child indicative that the death of the first child was *not* unnatural. Recently, the Court of Appeal quashed the conviction of Ms Angela Cannings for the murder of two of her infant children, three of whom had suffered 'an acute or apparent life-threatening event', on the grounds that the presumption of innocence in criminal cases requires statistical evidence to be analysed from a 'starting point' of the accused's innocence with respect to each death. The Court noted in its 2004 ruling:

> It would indeed have been an affront to common sense to treat the deaths of the three children . . . as isolated incidents, entirely compartmentalised from each other . . . Nevertheless a degree of caution was necessary to avoid what might otherwise have been the hidden trap of taking the wrong starting point. If, for example, at post mortem it was positively established that Matthew's death had resulted from natural causes, the situation reverted to precisely where it stood before he died. The concerns which would have arisen as a result of his death – as the third in the sequence – would have been dissipated. There would have been a positive innocent explanation for the death, which would no longer be a SIDS, and might help to confirm that the earlier deaths were indeed natural deaths.

From this starting point, the subsequent death of another child is not consistent with an explanation that the mother suffered from Munchausen's by Proxy (whereby the mother seeks attention by causing harm to her child and creating a need for medical treatment), and is not a basis from which her guilt may be inferred.

The probative value of statistical evidence is doubtful not only when it is tendered to prove that a crime was committed, but also when it is used to prove the identity of the criminal. Glanville Williams illustrates the irrelevance of statistics in proving the identity of an actor, by noting (in the *Criminal Law Review*) that: 'If the Blue Bus Co. has far more buses on the road than the Red Bus Co., this is no reason in law for assuming that the plaintiff was knocked down by a blue bus rather than a red bus.' Statistical reasoning unfortunately featured in the proof of identity in an American case where the accused was convicted of starting fires in four different properties. The likelihood of the accused being the arsonist in all four cases was inferred from the improbability of there being four fires. Roger Ede reports in *The Times* (3 February 2004) that the 'probability of four fires was calculated as the fourth power of the probability of one fire' without 'relating [the probability] to the specifics of the group being considered', which, in this case, was that of comparable properties.

Reliance upon even the most common form of statistical evidence to prove identity – DNA evidence – must be cautioned against. At a forensic science conference in 2003, Angela Gallop described the case of an accused charged with burglary. Whilst the forensic evidence estimated that his DNA profile, which matched the sample obtained from the scene, occurred in 1 in 37 million people, a police re-test using upgraded methods of analysis and a larger sample area 'demonstrated emphatically that he could not be the source of that DNA' (Roger Ede, 'Wrongful Convictions put Forensic Science in the Dock', *The Times*, 3 February 2004). For these reasons, the law is hesitant to harness statistical evidence in drawing inferences as to either that an act was committed or that the identity of an actor is that of the accused. Articulating the sentiment commonly held by judges, Charles Darwin, in a letter to Charles Lyell, commented that: 'I should think your remarks were very just about mathematicians not being better enabled to judge of probabilities, than other men of common sense.'

However, in particular types of litigation, statistical evidence is decisive. This is so where the statistics assess the probability not of whether an individual engaged in particular conduct, but of whether an individual's conduct caused the consequences alleged.

Proving causation in corporate manslaughter prosecutions: statistical evidence

The type of litigation in which statistics star is that labelled in the United States as 'toxic tort litigation', where corporations are held liable for the damage done to the health and livelihood of those affected by the corporations' pollution. Increased toxic tort litigation is looming on the UK legal horizon, with proposals afoot for reform of the law on corporate manslaughter. Pursuant to these reforms, corporations may be held criminally responsible for 'a gross management failure on its part' where that management failure 'caused a person's death'. A 'management failure' would be defined as 'a breach of its legal requirements to secure health and safety'. It is feasible that a 'management failure' may include a management policy that permitted the corporation's emission of pollution in its operations. Consequently, it is possible that the introduction of the reforms to the law on corporate manslaughter may make corporations criminally responsible when the toxic waste emitted has the effect of causing another person's death.

In anticipation of these proposals, a glance at a previous UK decision reveals the difficulties that prosecuting authorities may encounter in proving that the acts of a company were causative of a person's death. These difficulties of proof stem from the courts' reliance upon scientific data, whose conclusions must be proved to a scientific standard of proof, in order to satisfy the legal standard of proof. A predecessor to this strand of corporate liability lay in the Atomic Energy Authority Act 1954. In 1993, the relatives of a child who died from lymphoma sued the British Nuclear Fuels company, under section 5(3) of this Act, for damages. At trial, the corporation's negligent conduct was not in dispute. The only issue, according to the Court, was whether 'the cause or a material contributory cause of Dorothy's death was ionising radiation emitted by the activities carried on at Sellafield in Cumbria by the Defendants or their predecessors'. Dorothy was not directly exposed to the radiation. Instead, the chain of causation was indirect. The plaintiffs alleged that Dorothy's father was so exposed, so that the radiation caused 'mutation in the spermatagonia which in turn, via paternal sperm, causes a predisposition' to lymphoma in the next generation. This hypothesis of causation was known as the 'Gardner hypothesis', after an epidemiological study carried out by Professor Gardner which posited the hypothesis. The study observed the incidence and nature of lymphoma in people under 25 in a specified area, West Cumbria, between the years of 1950 and 1985. The study then used the data to assess the causal connection between a father's exposure to radiation before a child's conception and a child's development of lymphoma.

During the trial, Professor McMahon identified the methodology by which scientists investigate the truth of hypotheses of causation like those of Professor Gardner:

> In the end, it must be recognised that the idea of cause is a probabilistic one. Rarely can we be certain that a causal relationship exists, but by assembling evidence from many different angles we may build a body of support sufficient to convince most reasonable people that it is more prudent to act as though the association were causal than to assume that it is not.

The Gardner study did not assess the *probability* of the causal association, but assessed the *improbability* of there being no association. The standard of probability at which epidemiological studies traditionally characterize scientific evidence as 'significant', and indicative that the association between an alleged cause and its effect was not from chance alone, is when the evidence reveals a 95 per cent 'P value'. This means that a study is indicative of that association only when the occurrence of the reverse hypothesis (of non-causation) was 'improbable to the extent that chance would produce the result one time in 20'. Thus, the Gardner study had probative value to indicate that Dorothy's development of lymphoma was associated with the exposure of Dorothy's father to radiation only if the study revealed that association to a scientific standard of significance of 95 per cent. Evidently, the 'reasonable people' to whom science makes reference in determining the truth of a hypothesis of causation are, as a group, more sceptical than the reasonable person riding the Clapham omnibus to whom the law turns to determine causation. This disparity in standards of proof presented a problem to the trial judge: to which reasonable people ought the courts to defer when using scientific evidence to satisfy a legal standard of proof? Mr Justice French chose the reasonable and scientifically minded people. He concluded that: 'The fact that an epidemiologist or another scientist would not find an association and/or a cause to be established to his satisfaction is, of course, most helpful to a Judge but only within the limits imposed by their respective disciplines.'

The epidemiologist's discipline is limited by principles of scientific integrity that demand exactitude. Professor Carl Cranor (writing in *Science of the Law*, 5, 1993) explains that 'scientists want to be at least 95 percent sure [so] that they are not falsely adding to the stock of scientific knowledge when they report new discoveries or new statistical results'. Justice French adopted similar restraint and decided for the defendants: 'my conclusion is that though an arithmetically strong prima facie association is shown to exist, considerable reserve is necessary before placing reliance on it'. However, I would argue that the reasoning of Mr Justice French does not sit easily with the House of Lords in *Yorkshire Dale Steamship Co Ltd* v. *Minister of War Transport*. In that case, Lord Wright stated that: 'This choice of the real or efficient cause from out of the complex of the facts must be made by applying commonsense standards. Causation is to be understood as the man in the street, and not as either the scientist or the metaphysician, would understand it.' Just as the reasoning of judges in law cases does not affect the results of a scientist in the laboratory, concerns of scientific accuracy ought not to regulate the circumstances in which courts dispense justice. This is because the legal methodology, unlike the scientific, is concerned with the determination of the rights of the parties that appear before it.

Justice is dispensed on a case by case basis, so that the determination of criminal guilt or civil liability depends upon the facts of each case. The common law method of reasoning from precedent deems authoritative only the judicial decisions that relate to the law, but not the decisions concerning the facts of a case. Whilst the common law method requires courts of appeal to exercise caution when extending the stock of principles of liability to new categories of cases, the method does not require trial judges to exercise the precision of scientists when ascertaining the facts of the case. The law has its own standards of proof, to which precedence must be given. As noted, the criminal standard of proof beyond reasonable doubt compels a degree of certainty that exceeds the balancing exercise of the civil standard. Neither the civil nor criminal standards of proof are susceptible to statistical quantification, but are instead satisfied by reference to the explanatory power of competing hypotheses. The legal standard of proof involves assessments of rational probability, and not statistical probability, so that the criminal burden of proof beyond reasonable doubt is

satisfied when (according to Glanville Williams) 'the only reasonable explanation of the facts is that the defendant is guilty'. Thus, the element of causation
in the offence of corporate manslaughter is proved, to the criminal standard of
proof, when the scientific evidence indicates that the defendant corporation's
conduct constitutes the *only reasonable* explanation of the deceased's death. The
purpose of the scientific evidence is merely to highlight the reasonableness of a
hypothesis of causation, and to establish that, given the reasonableness of this
hypothesis, no other explanation is feasible. The degree of statistical probability (regarding the scientific truth of the hypothesis) that is required to establish
reasonableness is an issue for the trier of fact, having due regard to all the other
facts of the case, as well as the passenger on the Clapham omnibus. At this
point, scientists may perceive the courts as sacrificing a concern for the truth
to the hands of methodological slackness. Yet, the justification for the courts'
approach of reasonableness goes far beyond the unavailability of resources for
thorough scientific investigations. It lies in the role that the judicial process
serves in our society. In Dorothy's case, the judicial process serves to achieve
justice between the relatives of a child whose death was arguably caused by
radiation, and the company that emitted that radiation. In cases that may
be brought pursuant to the reforms to the laws of corporate manslaughter,
the judicial process serves to achieve justice by ascertaining whether or not a
corporation ought to have criminal sanctions imposed for its actions.

The courts accept or reject a hypothesis of causation for the purpose of either
resolving a civil dispute between two parties or ascertaining the criminal guilt
of a single accused. The courts do not accept hypotheses of causation in order to
advance scientific understanding, and scientific understanding does not depend
upon their findings. The focus of the courts' concern in these cases is narrower
than the focus of scientists. And for this reason, the courts are, and must be,
content to accept one contested version of reality as the closest conception of
truth to which the courts' grasp may reach.

Proving causation in corporate manslaughter prosecutions: the burden of dis-proof

The court's focus upon its function as an adjudicator for justice, and not as a
vehicle of scientific enquiry, has a second consequence: the court may be called
upon to pass judgment upon the issue of causation, even where the limits of

scientific knowledge render scientific evidence unavailable. This is the situation that occurred in the 1973 case of *McGhee* v. *National Coal Board*. The plaintiff, whose duties of employment included emptying brick kilns, was found to be suffering from dermatitis. He sued his employer for the breach of its duty 'to take reasonable care to provide adequate washing facilities including showers, soap and towels to enable men to remove dust from their bodies'. Whilst the employer admitted that it had breached its duty, the expert medical evidence 'could not do more than say that the failure to provide showers materially increased the chance, or risk, that dermatitis might set in'. Once more, the scientific evidence was inconclusive on the issue of causation. Regardless, the House of Lords ruled that the requirements of justice between the parties demanded a ruling in favour of the plaintiff. Lord Wilberforce reasoned that:

> In many cases, of which the present is typical, [the issue of causation] is impossible to prove, just because honest medical opinion cannot segregate the causes of an illness between compound causes. And if one asks which of the parties, the workman or the employers, should suffer from this inherent evidential difficulty, the answer as a matter of policy or justice should be that it is the creator of the risk who, ex hypothesi, must be taken to have foreseen the possibility of damage, who should bear its consequences.

In fulfilling its role as a mechanism for the just resolution of disputes, the House of Lords considered its departure from the strict 'logic' of proof justified. Lord Reid echoed the thoughts of Lord Wright in the *Yorkshire Dale Steamship* case, when stating that: 'The legal concept of causation is not based on logic or philosophy. It is based on the practical way in which the ordinary man's mind works in the everyday affairs of life.' However, in shifting to the defendant the burden of disproving the issue of causation, the House of Lords was enacting a fundamental tenet of Popper's system of logic: the House exposed the issue of causation to falsification.

To a lesser degree, the House of Lords followed the approach enunciated in *McGhee* in the recent 2003 asbestos case of *Fairchild* v. *Glenhaven Funeral Services*. In this case, it was settled that any cause of the plaintiff's mesothelioma other than the inhalation of asbestos dust at work could be effectively discounted. The difficulty arose because the plaintiff was exposed to the inhalation at two places of employment, and 'the current limits of human science' were unable to identify positively the extent to which the two employers individually caused the damage. As had been the case in *McGhee*, the House recognized the logical difficulty of requiring the plaintiff to verify the issue of causation with respect to each defendant in order to satisfy the burden of proof. Guided by notions of 'justice' and 'common sense', the House allowed that the issue of causation need not (to use Karl Popper's terminology) be 'conclusively decidable'. Lord Bingham declared the willingness of the Court: 'to treat the conduct of [both employers] in exposing [the plaintiff] to a risk to which he should not have been exposed as making a material contribution to the contracting by [the plaintiff] of a condition against which it was the duty of [the employers] to protect him'. It seems that the issue of causation was varied so that the plaintiff need only admit 'sufficient evidence to raise an issue as to the existence' of causation as a fact in issue. Whilst the plaintiff retains what is known as the 'evidential' burden of proof, the defendant is given the burden of disproving the remaining 'persuasive' burden of proof. Where the defendant fails to do so, the requirements of justice enable the court to find for the plaintiff.

To summarize, in disputes focusing on the liability of a corporation for tortious or criminal acts, two issues concerning the proof of causation arise. First, the admissibility of statistical evidence is governed by the legal, and not the scientific, standards of proof. Secondly, where scientific evidence is inconclusive or unavailable, and where 'the interests of justice' require, the courts exhibit a willingness to either shift to the defendant the burden of disproving causation or, upon the satisfaction by the plaintiff of an 'evidential burden' of proof, shift the remaining, and substantive, persuasive burden of proof.

The contrast between these two types of evidential issues highlights an interesting paradox in the symmetry of the structure of legal proof and Popper's *Logic of Scientific Discovery*. The courts' reliance upon the notion of reasonableness to determine the *standard* of proof makes Popper's formal theory of proof an inappropriate method of probability reasoning. Notions of reasonableness may justify the courts' reliance on scientific evidence that is not proved to the requisite degree of scientific certainty because the courts have a peculiar concern with the interests of justice. Yet, the intuitions of the rider of the Clapham omnibus do not constitute a 'probability premise' from which all of Popper's remaining 'probability conclusions' are formally derived.

On the other hand, in the cases of *McGhee* and *Fairchild* the courts rely on notions of reasonableness and the demands of justice to shift to the defendant corporation the burden of disproving causation. In so doing, the courts thereby subject the issue of causation to Popper's empirical structure of proof by falsification, and not verification. This method of proof, whereby facts in issue are left to be disproved by the party against whom the allegations are made, appears particularly suitable to cases determined by international tribunals. In the case of *Velásquez Rodríguez*, the Inter-American Court of Human Rights found (as detailed in *Human Rights Law Journal*, 9, 212, 1988) the state of Honduras responsible for the disappearance of Mr Rodríguez, and, thus, in violation of Articles 4, 5 and 7 of the American Convention of Human Rights, even though the complainant was unable to elicit evidence as to the State's involvement in his kidnapping:

> In contrast to domestic criminal law, in proceedings to determine human rights violations the State cannot rely on the defense that the complainant has failed to present evidence when it cannot be obtained without the State's cooperation . . . The facts reported in the petition whose pertinent parts have been transmitted to the government of the State in reference shall be presumed to be true if . . . the government has not provided the pertinent information.

The case of *Velásquez* is of further interest in that it highlights that international tribunals – unlike domestic courts, whose procedures are ultimately enforceable by a sovereign government – face unique issues of evidential proof. Such difficulties of proof in international tribunals are addressed in the following section.

Methods of proof: the Darwinian fact-finding explorations of international tribunals

The imperfections of the trial process as a mechanism of finding the truth are similarly prevalent where the jurisdiction of the courts concerns the truth of accusations of the most heinous of crimes, being war crimes and crimes against humanity. For this reason, the international criminal courts utilize the same criminal standard of proof beyond reasonable doubt.

However, the focus of international criminal courts differs from that of the courts that are a part of domestic and sovereign systems of government in two ways. First, a war crimes trial is an exercise in partial justice to the extent that it reminds us that the majority of war crimes go unpunished. This was a criticism in particular of the Yugoslavia Tribunal's decision to prosecute Dusko Tadić – a mere foot-soldier in the events of the Balkan crisis – simply because it did not have custody of a higher-ranking, more significant figure. It was argued that there were hundreds more like Tadić, and that there was little point in convicting one among them in what seemed to have been a mere lottery.

Secondly, the court's inability to try all cases is balanced by the fact that the focus of international criminal trials is broader than the case at hand. In a world in which a multitude of people may have become embroiled in war crimes, the punishment of each and every offender is not necessary to achieve respect for the rule of law, or to declare our disgust at the acts committed. Unlike domestic criminal and civil courts, where the sovereign state requires the courts to determine all disputes that arise between the state's subjects, international criminal courts have a commemorative potential; they can build, as Antonio Cassese has pointed out, an objective and impartial record of events. This was true of Nuremberg, it is true also in respect of the current trials in The Hague, and it appears to be especially true of the 'Extraordinary Chambers' that were recently established by the UN and the Royal Government of Cambodia to try those alleged to have committed atrocities during the rule of Pol Pot and the Khmer Rouge between 1975 and 1979. Naturally, we should recognize the tension between the production of history and the task of conducting a criminal trial. A criminal trial with its elaborate rules regarding relevance and inadmissibility of evidence, as well as its focus on the accused in court, can never provide a definitive and comprehensive record of history. International criminal tribunals are able to provide a coherent and judicially manageable account of tragic events, a 'judicial truth', in effect. But the painting of the fuller picture of history, through local initiatives such as truth commissions based on popular participation, must be left to those affected by the crimes. That having been said, we can rest assured that the materials collected by these international tribunals which have passed its strict rules of admissibility of evidence can contribute to the creation of objective accounts of events which will play an important role in fighting forgetting.

The amount of evidence that is available to each of these international tribunals varies significantly. Patricia Wald, a Judge of the International Criminal Tribunal for the former Yugoslavia (ICTY), sat on the trials of 'guards and lower level officers, such as the Omarska and Keraterm camps cases' and reported in an article in the *Washington University Journal of Law and Policy* that:

> Even in the most monstrous war crimes involving executions and massacres of thousands there may be no evidence of written orders to execute, bury, or rebury the victims, nor sure identification of the senior commanders who

actively planned, approved, or ordered the slaughter. The only documents in these trials may consist of coded excerpts from Bosnian–Serbian intercepted phone calls, often subject to varying interpretations.

In contrast, due to the assiduous record-keeping habits of the Khmer Rouge, and the diligent explorations of historians during the 20 years since the atrocities, the Extraordinary Chambers looks set to receive a deluge of documentary and other evidence when it opens its doors. Ben Kiernan, founder of the Yale Cambodian Genocide Program, reported that, in 1996, the Program obtained a 50 000-page trove of documents produced by the former Khmer Rouge regime's security police, the Santebal. The Program has since distilled the dossier and published on line 'over 19 000 biographic entries on Khmer Rouge officials and their victims, 3000 bibliographic records on sources of information about that era, and over 6000 photographs, documents, translations, and maps'. Hauntingly, the database contains over 5000 images of Khmer Rouge victims who were photographed by Khmer Rouge personnel at the infamous Tuol Sleng prison and torture centre.

A consideration of the experiences of the ICTY, and the expected complexities of the Extraordinary Chambers, reveals that, regardless of whether an international tribunal is admitted, an abundance or a scarcity of evidence of war crimes, difficulties of proof inhere. As the commemoration of many victims rests upon the guilty verdict for a single accused, the focus of international criminal courts is broader than that of domestic courts. This broader focus mandates a broader evidential approach than that of domestic courts. The methods of proof that the international tribunals adopt are invariably a hybrid of practices imported from domestic legal systems. This is because the tribunals must be seen to be impartial and not a mere instrument of any particular country. One way that this is achieved is to amalgamate the rules of procedure of many countries. And, as the procedures of the ICTY, and the expected complexities of the Extraordinary Chambers, reveal, this amalgamation creates unique barriers to the international courts' elicitation of the truth.

Common law and civil law traditions: the ICTY

Where documentary evidence is lacking, international tribunals rely on the personal testimony of victims who survive massacres. During its period of operation, the ICTY has brought before it nearly 1000 victim-witnesses to provide testimony. The procedure by which the ICTY receives the testimony of witnesses is detailed in its Rules of Procedure and Evidence, which imposes a system of procedure that amalgamates the rules of the common law and civil law systems. Judge Wald has described the difficulties that this fusion causes for the effective elicitation of evidence:

> the bulk of defense counsel are Balkan-trained lawyers and are typically not experienced at cross-examination . . . They sometimes argue with or even criticize the witnesses. They also go off on tangents that are not always relevant to their case. The Tribunal is now operating training courses for appointed lawyers, but, candidly, it is not easy to acculturate lawyers in a wholly new legal system in a few days of lectures or even simulated exercises . . . In sum, I came away from the two lengthy trials in which I have participated thinking that the potential of cross-examination by defense counsel in the search for truth has not been realized.

International and domestic traditions: the Khmer Rouge tribunal

On the other hand, where an international tribunal is faced with masses of documentary evidence that implicates a number of people and details a plethora of atrocities, the primary issue that faces the tribunal is: Which crimes ought to be prosecuted? The nature of mass atrocities, such as those committed in Cambodia and Yugoslavia, make it nearly impossible for the prosecutor to avoid making broader strategic choices when it comes to deciding whom to prosecute. This is a scenario that the USA currently faces. In Iraq, the coalition forces are reported to have captured 7000 combatants, all of whom are being detained until a US military tribunal, composed of 3 US military officers, makes a determination as to their status as prisoners of war.

The June 2003 agreement between the Senior Minister in the Royal Government of Cambodia, Sok An, and the UN Legal Counsel, Hans Correll, created a tribunal to try those suspected of Khmer Rouge atrocities that is part-international and part-domestic. The Extraordinary Chambers will be established within the existing Cambodian court structure, whilst the composition of the trial and appeal courts of the Chambers, as well as the prosecutors, will be a mixture of Cambodian and foreign lawyers. The decision as to which suspects to prosecute will be a joint decision of the foreign and Cambodian co-prosecutors. The potential for discord in the making of this assessment was apparent to the drafters of the agreement. Article 20 of the agreement details the procedure by which disputes as to a decision to prosecute may be resolved, and provides that:

> The Co-Prosecutors shall submit written statements of facts and the reasons for their different positions to the Director of the Office of Administration. The difference shall be settled forthwith by a Pre-Trial Chamber of five judges, three appointed by the Supreme Council of the Magistracy, with one as President, and two appointed by the Supreme Council of the Magistracy upon nomination by the Secretary-General of the United Nations.

There is potential for the decision-making procedures of the Extraordinary Chambers to frustrate the process by which every suspect is brought to justice. However, the Extraordinary Chambers, like other international criminal courts, is a call to responsibility for persons guilty of (in the words of the preamble to the Statute of the International Criminal Court) 'the most serious crimes of concern to the international community as a whole'. In this respect it takes seriously the words of Justice Robert Jackson, Chief Prosecutor at Nuremberg, who famously said that letting major war criminals live undisturbed to write their 'memoirs' in peace 'would mock the dead and make cynics of the living'.

The Chambers present themselves as the mechanism for providing a public demonstration of justice. The act of punishing particular individuals – whether the leaders, or star generals, or foot-soldiers – becomes an instrument through which individual accountability for massive human rights violations is increasingly internalized as part of the fabric of our international society. At the same time, it is a method by which we put a stop to the culture of impunity that has taken hold at the international level. Former US Secretary of State Warren Christopher suggested in the context of the Balkan crisis that '[b]old tyrants and fearful minorities are watching to see whether ethnic cleansing is a policy the world will tolerate'.

If the function of a trial in international tribunals is first and foremost a proclamation that a certain conduct is unacceptable to the world community, then justice need not be wreaked by ensuring that every person who is suspected of crimes of concern is brought to account. The purpose of the international tribunals mandates a different approach from that of domestic tribunals. So, if the cosmopolitan procedures of international tribunals enable courts to make this proclamation, and have it accepted by the world's people as the proclamation of a legitimate judicial authority, then it may be said that the evidential headaches that the procedures cause its lawyers are bearable.

Conclusion

I have highlighted three areas in which the judicial search for truth is imperfect. Concerns for victims of domestic violence are balanced against the accused's evidential right to a presumption of innocence. A corporation's reliance upon the stringent scientific standard of proof is sacrificed to the court's focus upon the guilt or innocence of the accused corporation. And finally, the bringing to trial of every suspected war criminal is relinquished in return for the commemorative value to wider societies of a declaration of the guilt of individual war criminals.

Whether at a domestic level or at an international level, the process of a trial is an imperfect mechanism for finding the truth. By using different rules of evidence and standards of proof depending on the nature of the task, it is possible to devise a process which, whilst not perfect, comes as near as is practically possible to fulfilling the important role of the judicial process in society, which is to achieve justice between the parties. And it is the court's adjudication between the competing verities posited by the advocates, a judgment reached by observing the legal evidence which is admitted during the 'tight-rope walk' that is the courtroom battle, that makes the verdicts of the judicial process as near to 'truth' as possible. For as Einstein reflected: 'Knowledge cannot spring from experience alone but only from the comparison of the inventions of the intellect with observed fact.'

FURTHER READING

Balding, D. and Donnelly, P. , 'The Prosecutor's Fallacy and DNA Evidence', *Criminal Law Review* 711, 1994.

Cohen, L. Jonathan, *The Probable and the Provable*, Clarendon Library of Logic and Philosophy, Oxford: Clarendon Press, 1977.

Epstein, D., 'Effective Intervention in Domestic Violence Cases: Rethinking the Roles of Prosecutors, Judges, and the Court System' *Yale Journal of Law and Feminism*, 11, 3, 1999.

Kuhn, T., *The Structure of Scientific Revolutions*, Chicago: The University of Chicago Press, 1970.

Popper, K., *The Logic of Scientific Discovery*, London: Unwin Hyman, 1959.

Wald, P. M. 'The International Criminal Tribunal for the Former Yugoslavia Comes of Age: Some Observations on Day-To-Day Dilemmas of an International Court', *Washington University Journal of Law and Policy*, 87, 5, 2001.

Williams, Glanville, 'The Mathematics of Proof – I', *Criminal Law Review*, 297, 1979.

8 Evidence for Religious Faith: A Red Herring

KAREN ARMSTRONG

Evidence for the existence of God

When I was about eight years old, I had to learn this definition of God from the Roman Catholic Catechism: 'God is the Supreme Spirit who alone exists of Himself and is infinite in all perfections.' At the time, it didn't mean much to me, but later it became very troubling, because I could find no evidence to support the belief in this Supreme Being, who, according to the Bible, had created Heaven and Earth. I am highly conscious that I have been invited to give this lecture by Darwin College, and it is a matter of common knowledge that Darwin's Theory of Evolution shook what is presumed to be the traditional belief in the Genesis account of a seven-day creation. What were usually known as the Five Proofs for the existence of God did not seem convincing to me as a teenager, and became still less compelling in adult life. Then there was the matter of Jesus and the redemption. It seemed clear that Jesus had been a historical figure, crucified by the Romans in about 30 CE, but how did anybody *know* that he was divine? Modern New Testament scholarship has even disabused us of the old belief that he claimed to be the Son of God as described in the Nicene Creed. And so gradually, in the absence of incontrovertible evidence, my faith in God and Christianity died away, and for many years I had no time at all for religion.

I was quite correct in doubting the evidence for the truths of my faith. There simply is no evidence that proves, empirically and logically, the existence of God or the divinity of Christ. There is no proof for the existence of Nirvana, Brahman or the Tao, either. These realities are not capable of rational demonstration, because they lie beyond the senses and beyond mundane experience. What I had not appreciated was that leading sages, prophets, philosophers and theologians had always been aware of this, and that major Jewish, Christian and Muslim thinkers would have taken severe exception to the definition that I quoted above. First, they would have unanimously averred, God is not the 'Supreme Being', because that would indicate that 'he' – a ridiculous pronoun – is simply a being like ourselves, writ large, with likes and dislikes similar to our own, and that was unacceptable. Second, God did not 'exist' because our notion of 'existence' is far too limited to apply to God, who does not exist in the same way as this book, another human being, or Australia, and he is not an unseen reality, like the atom, the existence of which can be demonstrated mathematically in a laboratory. When we are speaking about God, we are talking about a different kind of reality altogether. Third, they would have pointed out that the Catechism description was itself incoherent: if God was 'infinite' how was it possible to sum him up in a one-sentence 'definition' – a word that literally means 'to set limits upon'. Finally, the definition assumes that God is male, and what we call 'God' transcends gender, as it goes beyond any other human category.

Indeed, one of the reasons why the Greek Orthodox theologians formulated the doctrine of the Trinity in the fourth century was to remind Christians that they could not talk about God as though he were a simple Personality. Long before modern atheists had denied God's existence, monotheistic theologians had asserted that, strictly speaking, there was nothing out there. Like many of the Muslim theologians, Maimonides constructed a contemplative exercise. First you should say to yourself: 'God exists.' But that is incorrect, for the reasons I have already given; next you should say 'God does not exist', but that is not quite right either. Finally, you are reduced to saying: 'God does not not exist.' This was not just a pointless conundrum; it was a meditation on the inadequacy of language in the face of transcendence: by allowing these propositions to reveal themselves as inadequate, you came to realize, at a level deeper than the cerebral, that when we are speaking about the divine we are at the limits of what words and thoughts can do. Maimonides also said that it was better to call God 'Nothing', because God was not another being. Jews refrain from speaking God's name, in the same way as Muslims forbid any visual representation of the divine, as a reminder that any human expression of God is bound to be so limited as to be potentially blasphemous.

The fifth-century theologian who wrote under the pseudonym 'Denys the Areopagite', whose work has near-canonical status in the Orthodox Church, pointed out tirelessly that what the Bible reveals about God – his goodness, paternity, wisdom, omnipotence and so forth – was not God itself. If we really want to understand what we mean by 'God' we must go on to deny those attributes and names. Thus we must say that it is both 'God' and 'not-God', 'good' and 'not-good'. The shock of exploring this paradox, a process that includes both knowing and unknowing, will lift us above the world of mundane ideas to the indescribable reality itself. Thus, we can start by saying that: 'Of him there is understanding, reason, knowledge, touch, perception, imagination, name and many other things. But he is not understood, nothing can be said of him, he cannot be named. He is not one of the things that exist.' Reading the Bible, therefore, does not give you facts about God, but should be a paradoxical discipline. The method described by Denys is a method to prevent us from thinking logically about God. We even have to leave our denial of God's attributes behind. It is certainly pointless to try to find evidence for the divine. You cannot prove the existence of nothing.

Two objections are likely to spring to mind. First, if God – or Nirvana, Brahman or the Tao – do not exist, what is the point of religion (a point which I address below)? Second, religious people are constantly defining the truths of their faith, and forcing others to do the same. This, I believe, is what the Buddhists would call 'unskilful' religion. Because we place such high value on rational understanding in our scientific culture, we have lost sight of the fact that the transcendence *is* transcendence: it goes beyond logic and reason in the same way as music or poetry. You cannot find evidence for the 'truth' of a sonnet by Rilke or a Beethoven Quartet. Religion should be regarded as an art form, rather than an exact science or philosophy. As a means of illustrating how far we in the West have travelled from Denys' apophatic or silent theology, consider the following two theological terms. The first is *theoria*: in Greek, this was contemplation; our 'theory' is a mental construct. Second is the word *dogma*. For Greeks like Denys, *dogma* was everything about religion that you could not define, could not speak about, a wisdom that could only be learned by long contemplative discipline, and it was ineffable. For us in the West, 'dogma' and 'dogmatic' mean almost precisely the opposite. How did we come to this?

Mythos and logos

In pre-modern society, it was generally recognized that there were two ways of arriving at truth. Plato called them *mythos* and *logos*. They were both regarded as indispensable, and as complementary, and each had its special area of competence. Myth was not concerned with practical matters, but with meaning, which is beyond the reach of science. Human beings are so constructed that they fall very easily into despair, in a way that other animals do not, when they find no significance in their lives. Myth was rooted in the unconscious mind, was an early form of psychology, and could not be demonstrated by rational proof. It was also connected with mysticism and religion. It only made sense when embodied in rituals and ceremonies that worked aesthetically upon worshippers, like a great work of theatre, enabling the participants to apprehend the deeper currents of their existence. *Logos* was equally important. It was the rational, pragmatic and scientific thought that enabled men and women to function effectively in the world. Unlike myth, *logos* must relate exactly to facts and correspond to external realities; it must work efficiently in the mundane world. We use this logical, discursive reasoning when we have to make things happen, get something done, or persuade other people to adopt a particular course of action. Unlike myth, which looks back to the beginnings, *logos* forges ahead and tries to find something new, achieve a greater control over our environment, discover something fresh, and invent something novel.

Both myth and *logos* had their limitations. Myth could not help you to organize your society, solve a mathematical equation, or formulate a viable economic policy. It gave you no information about the external world. But equally *logos* could not assuage human pain and sorrow, or answer questions about the ultimate value of human life. A scientist could make things work more efficiently and discover wonderful new facts about the physical Universe, but he could not explain the meaning of life. If your child dies or you witness a terrible natural disaster, you do not want a logical discourse to explain what has happened. You need the kind of comfort that was traditionally provided by myth.

It was usually considered inadvisable to confuse *mythos* and *logos*, although accidents did and continue to happen. Religion, God and the like clearly belong to the realm of myth, and if you try to apply the rules of *logos* to the first chapter of Genesis, you get bad science and bad religion. Nevertheless, people did try to apply the truths of logic and reason to their faith, and to find some rational evidence for their beliefs in order to make their faith cohere with the dictates of reason.

Islamic philosophy and mysticism

This experiment began in the Islamic empire during the eighth and ninth centuries, when the Muslims had encountered the writings of Plato, Aristotle and the Greek scientists and mathematicians. The encounter inspired a cultural florescence, a cross between our European Renaissance and the Enlightenment, during which more scientific discoveries were made by Muslim scientists than in the whole of history hitherto. They called their movement 'Falsafah' ('philosophy'). Some of these Faylasufs tried to make the religion of the Koran a rational faith and used Aristotle's demonstration of the Unmoved Mover to prove the existence of Allah. Aristotle had argued – not very effectively – that in a rational universe everything had a cause; everything alive was mobile and motion was always initiated by something external. There must therefore be an Unmoved Mover to start the ball rolling. The Muslim philosophers applied this series of 'proofs' to the God of the Koran, but their God bore no resemblance to Allah. There is nothing religious in Aristotle's God, who seems scarcely aware that the human race exists. There is no point in praying to him, he does not care how humans behave, and they will never be able to know anything about him. Nevertheless the Faylasufs, who were also interested in the political science of Plato's *Republic*, did some important work. Jews living in the Islamic empire also became Faylasufs, and set about proving the existence of the God of the Bible.

Interestingly, the Greek Orthodox Christians wanted nothing to do with this project. It seems as though they understood the limitations of Greek metaphysics. They might have said that trying to prove the existence of God by means of reason was about as useful as eating soup with a fork. A fork is a perfectly acceptable implement for some kinds of food, but hopeless for soup. In the same way, *logos* was indispensable for mathematics, medicine and natural science, but entirely inappropriate for the study of God. You could not be a theologian unless you were engaged in full-time contemplation and were fully involved in the liturgy. From the very beginning, all the really important Orthodox theologians were deeply spiritual men: one need think only of Athanasius of Alexandria who opposed Arius at the Council of Nicaea in 325 CE, the Cappadocian theologians – Gregory of Nyssa, his brother Basil of Caesarea, and their friend Gregory of Nazianzus – who formulated the doctrine of the Trinity in the late fourth century; Denys the Areopagite; and Maximus the Confessor

(580–662 CE), who is known as the founder of Byzantine Christology. The Greeks developed new criteria for theology. Any valid statement about God had to have two characteristics: first, it must be apophatic – that is, it must induce a state of silent awe; second, it must be paradoxical, to remind Christians that God could never be made to fit neatly into a human system, however august. After a while, both Jews and Muslims lost their enthusiasm for Falsafah, which remained a minority pursuit.

A pivotal figure of this development in the Islamic world was the eleventh-century theologian al-Ghazzali, who is often said to be the most important Muslim who ever lived, after the Prophet Muhammad. He was such an outstanding scholar that, at the age of thirty-three, he was appointed director of the prestigious Nizamiyyah madrasah in Baghdad. He had a restless temperament, which made him struggle with the truth like a terrier, worrying problems to death, and refusing to be content with an easy conventional solution. He noted that Muslims in his day sought religious certainty in four ways: Shiites revered the memory of divinely inspired Imams, who could lead them infallibly to truth; rational theologians and Faylasufs claimed that God was accessible to human reason; and Sufis followed a mystical path. But how could their claims be verified objectively? The reality that we call 'God' cannot be tested empirically, so how could Muslims be sure that they were not entirely deluded? The rational proofs failed to meet al-Ghazzali's strict standards. In his view, the Faylasufs were irrational and unphilosophical because they sought knowledge that lay beyond the capacity of the mind and could not be verified by the senses.

But where did this leave the honest seeker? The strain of his quest caused al-Ghazzali such personal distress that he had a breakdown. He found himself unable to swallow, felt overwhelmed by a weight of despair, and finally found that he was entirely unable to speak. The doctors rightly diagnosed a deep-rooted conflict and told him that until he was delivered from his buried anxiety, he would never recover. Al-Ghazzali resigned his academic post, and went off to join the Sufis and lived for ten years in Jerusalem practising the yoga-like exercises of concentration and breathing that have been developed all over the world as a means of gaining an experience of transcendence. Eventually al-Ghazzali decided that God is not an external, objectified being, whose existence can be proved by *logos*. Science could neither prove nor disprove his existence The only way to achieve religious certainty was by religious experience, which helped us to cultivate a different way of seeing. This proved to be a watershed.

The Koran had very little time for theological speculation, which it called *zannah* ('self-indulgent guesswork'); it seemed outrageous that people should quarrel over metaphysical concepts, such as the divinity of Jesus, that nobody could prove one way or the other. Islam, like Judaism, was a religion of practice rather than of belief. The Faylasufs had always been a tiny minority and never gained much support among the rank and file, who found their arguments odd, artificial and irrelevant. Al-Ghazzali was thus far more in tune with mainstream Muslim opinion than the rationalist philosophers who wanted to prove God's existence. The Koran made it quite clear that you could only talk about God in signs and symbols; Muslims knew instinctively that God and religion belonged to the realm of *mythos* rather than *logos*. Al-Ghazzali was regarded as the leading Muslim thinker of his day and, after his attack on Falsafah, Muslims abandoned it with an almost audible sigh of relief. Never again would Muslims make the facile assumption that God was a being whose existence could be proved like that of any other object in the material world. Henceforth, Muslim philosophy became inseparable from spirituality and a more mystical discussion of God, in the same way as the Greek Orthodox Church had. Until the nineteenth century, Sufism, which before al-Ghazzali had been a fringe movement, coloured the religion of most Muslims.

Genesis and myth

Jews came to the same conclusion as al-Ghazzali and his followers, but by a different route. They found that, when they were assailed by tragedy and cruelty, the remote God of the Philosophers was no use to them. Instead, increasingly, they turned to the *mythos* of the Kabbalah. An interesting and crucial figure was the sixteenth-century mystic Isaac Luria, who developed an entirely new creation myth that bore no relation to the account in Genesis. It is worth pausing here to consider the role of cosmogony in religion, because this has become such a problematic issue for many Christians. People used to see the marvellous design of our world as incontrovertible evidence of a Divine Creator. However, since the discoveries of Darwin and other scientists, who find no room for God in their vision of the Universe, many have felt their faith to be in jeopardy.

All religious people have developed a mythology about the origins of life, but this was not intended to give factual, historical or scientific information.

People know this is impossible, since nobody was around to witness these extraordinary events. The *Rig Veda* suggests that not even God, 'the overseer of the highest heaven', knows the secrets of creation. There could be no definitive answer. In the Bible, for example, the editors of Genesis have juxtaposed two entirely different and mutually exclusive creation myths, and placed them side-by-side. Other psalmists, sages and prophets referred to older myths, which imagined Yahweh bringing an ordered cosmos into being by fighting a dragon or a sea monster, emblems of formlessness and chaos, like many other Middle Eastern gods. Creation myths were primarily therapeutic. They were usually recited in a ritual contest, where the gods' primordial struggle with the forces of chaos was sometimes represented dramatically. It was hoped that worshippers would be able to tap into this immense creativity and power. Thus a cosmogony was usually recited at dangerously transitional moments, when there was always the possibility of a lapse back into chaos and disorder and an infusion of divine strength was required: at New Year, for example, or at the coronation of a new king. Cosmogonies would also be recited in a private setting – at a sickbed, for example – or at the beginning of a new enterprise, such as building a new house or a boat, when setting out on an expedition, or when founding a new settlement. People probably thought that these divine events had, in some sense, happened at one time, but *mythos* was never intended to be understood literally.

The creation story in the first chapter of Genesis was also primarily therapeutic. It was written by a priestly author during the Jews' exile to Babylon, as a polemic against Babylonian religion. Even though the Jews had been defeated by the Babylonians, who had laid waste their temple, their God was far superior to Marduk, the local creator god, whose exploits were celebrated in an elaborate liturgy, which we know deeply troubled the Jewish exiles. Yahweh did not have to fight a sea monster; Leviathan was simply one of his creatures. Marduk's victory had to be repeated every year, in order to prevent the slide back to primal chaos, but Yahweh's creation was effortless, and achieved once and for all. On the seventh day, he could rest from his labours, his work complete. But, however consoling this myth, it was not the last word on the subject. Jews continued to refer to the older myths of divine combat, as we see in the prophecy of Second Isaiah, who preached at the very end of the Babylonian exile and speaks of Yahweh fighting a monster called Rahab in order to bring the world into being.

FIGURE 8.1. Seated Sakyamuni Buddha from Ch'ungung-ni, tenth century (iron);
National Museum, Seoul, Korea.

FIGURE 8.2. Page from the *Mishneh Torah*, systematic code of Jewish law written by
Maimonides (1135–1204) in 1180, *c.* 1351 (vellum) by Jewish School (fourteenth
century); National Library, Jerusalem, Israel. Lauros / Giraudon.

Thus, when Isaac Luria put forward his new creation myth, some two millennia later, nobody was disturbed. Kabbalists had long interpreted the first chapter of Genesis symbolically, seeing every single word as an esoteric reference to the hidden life of the ineffable God. Luria's myth was quite different from the orderly account of Genesis. This creation is erratic and unpredictable, consisting of explosions, mistakes and false starts. A primary disaster means that sparks of the divine light have been trapped in the material world, and everything is in the wrong place. Today Luria would be severely castigated for contradicting the Bible, but at that time Lurianic Kabbalah became a mass movement throughout the Jewish world. It was able to do so not because it was scientifically verifiable, but because it reflected the conditions of the Jewish people's existence. Jews had recently been ejected from Spain by the Catholic monarchs Ferdinand and Isabella. During the fifteenth century, Jews had been expelled from one region, one city of Europe, after another. Consequently, they recognized the catastrophic cosmos of Isaac Luria, who taught them that exile was a fundamental law of existence. Even God was in exile from part of Itself, so in their exile Jews were not marginal outcasts, but central actors in the redemptive process. Luria's new myth showed that the careful observance of the commandments of the Torah, the Law of Moses, could end the divine exile, restore the Jewish people to the Promised Land, and the rest of humanity to its rightful state. Luria's myth was therapeutic to the traumatized Jewish people. Today, some Jews, such as the American scholar Richard Rubinstein in his celebrated book *After Auschwitz*, find that this is the only way they can think about God after the Holocaust: the impotent, wounded deity, who is not in control of his creation, makes more sense to them than the serene, all-powerful creator of Genesis.

Rationalism and belief

Jews and Muslims, therefore, withdrew from the attempt to find rational evidence for God, and resorted to a more mystical, mythical faith. It is, therefore, ironic, that, just as this happened, Western Christians became enthusiastic about the possibilities of theological *logos*. In the sixteenth century, they began their scientific revolution that would change the world forever, in many ways for the better. By the nineteenth century, *logos* had achieved such spectacular results that *mythos* was discredited, and instead of seeing the two as complementary, people began to see reason as the sole guide to truth. They also began to read their Bibles in a wholly literal manner, as though it were a holy encyclopaedia that gave factual information about God. We see this in the seventeenth-century British scientist Sir Isaac Newton, who developed the rigorous scientific methods of experimentation and deduction more completely than any of his predecessors. But his total immersion in the world of *logos* made it impossible for Newton to appreciate that other, more intuitive forms of perception might also offer human beings a form of truth. In his view, mythology and mystery were primitive and barbaric ways of thought. ' 'Tis the temper of the hot and superstitious part of mankind in matters of religion', he wrote irritably, 'ever to be fond of mysteries and for that reason to like best what they understand least'.

It was in the eighteenth century that the concept of faith became fused with the notion of belief. Hitherto faith had meant trust and commitment, as when we say that we have faith *in* a person or an ideal. The word *credo*, now translated as 'I believe', was probably derived from the words *cor do*: 'I give my heart.' It did *not* mean an intellectual assent to a set of creedal propositions. Similarly the Middle English 'beleven' meant 'to love'. In the New Testament, the Greek *pisteuo*, often translated 'I believe', meant 'I commit myself.' Thus, Theodore, Bishop of Mopsuestia in Cilicia from 392 to 428, explained to his converts: 'When you say "I engage myself" [*pisteuo*] before God, you show that you will remain steadfastly with him, that you will never separate yourself from him and that you will think it higher than anything else to be and to live with him and to conduct yourself in a way that is in harmony with his commandments' (Homily 13:14). He did not mention their submission to any metaphysical orthodox belief; faith meant emotional rather than intellectual commitment. It was not until after the European Enlightenment that Western Christians began to equate faith with belief. When Luther, for example, spoke of justification by faith, he did not mean that Christians would be saved by their conformity with the Creed but by their complete trust in Jesus.

Until the modern period, therefore, Christians had quite a different conception of faith. They would have found our current obsession with doctrinal orthodoxy very peculiar. Today people often translate St Anselm of Canterbury's dictum 'credo ut intellegam' as 'I believe in order that I may understand.' The implication is that one should submit to the orthodox Creed, however incomprehensible and offensive to reason it may seem. This act of intellectual surrender will be rewarded by true understanding. In fact, however, St Anselm probably meant 'I commit myself in order that I may understand', which perfectly expresses the older and more authentic Christian attitude.

In the modern world, we tend to assume that, before we can start to live a religious life, we should first convince ourselves of the evidence for God's existence. This is a good rational principle: first establish your facts, and then seek to apply them. There is not much point, after all, in going to the immense trouble of behaving like a Christian or a Muslim if there is nothing out there. But the masters of the spiritual life in all the major traditions would say that this is to put the cart before the horse. Religion is not about believing things; it is not primarily about thinking. It is about doing things, behaving in a way that changes you. What St Anselm meant was that an aspirant should first commit himself to the Christian life, and then he would begin to understand what it was all about.

The Buddha makes this very clear. He always refused to define Nirvana, because it was transcendent and inexpressible. There was no point in asking for evidence that Nirvana existed, because none was available. But if his monks adopted the Buddhist way of life – speaking the truth at all times, behaving with compassion and equanimity to all, and practising the disciplines of yoga – then they would have intimations of Nirvana, even though they would never be able to express this logically. One of his monks was of a philosophical turn of mind, and kept pestering the Buddha about the evidence for his faith, to the detriment of his yoga and ethical practice: Who had created the world? Did the gods exist? And was the world created in time or had it always existed? The Buddha told him that he was like a man who had been shot with an arrow, but who refused to have any medical treatment until he found out the name of his assailant and what village he came from. He would die before he got this perfectly useless information.

Religion as practice

Today we often describe religious people as believers, as though believing is the main thing that they do. But most of the major world religions are no more interested in metaphysics than was the Buddha. Judaism and Islam are both religions of orthopraxy, of right practice rather than right belief. The Koran indeed is highly sceptical about theological speculation. It is appalled that religious people argue about matters such as the divinity of Christ. The word *kafir*, usually translated 'unbeliever', actually means 'one who is ungrateful to God'. Like the Jewish prophets, the Koran preaches the importance of right action and social justice. There is very little doctrine in the gospels. Jesus spends no time discussing the Trinity, the Incarnation or Original Sin, but tells people how they should behave.

Religion can thus be defined as an ethical alchemy. As al-Ghazzali discovered, the only evidence that our faith is true will come from our experience of living in a religious manner. People have discovered, by trial and error, that living in a certain way brings an enhanced sense of humanity, wakens hitherto dormant qualities, and gives us intimations of a transcendent dimension – a transcendence that has been interpreted in many ways – as God, Nirvana, Brahman and the Tao; it has sometimes been seen as supernatural, to others it is simply a part of our human nature. But, however it has been defined, transcendence has been a fact of human life. A peculiar characteristic of the human mind is that it has the ability to have experiences and conceive ideas that it cannot wholly express or understand. Indeed, we seek out such experiences. We are beings that seek ecstasy, a sense that we inhabit our humanity more fully than usual. If we no longer find ecstasy in a synagogue, mosque, temple or church, we seek it in art, music, poetry, sex, dance, pop, football or skiing. Unless we have our share of such experience, we do not feel fully alive.

But we have to work to achieve ecstasy. If you walk into an art gallery, entirely ignorant of the tradition of Western painting, you are not going to understand Matisse or Monet. It takes time, effort and skill to read poetry. Similarly, to achieve religious ecstasy, we have to cultivate a special sensitivity, in the same way as we cultivate an aesthetic sense. As al-Ghazzali put it, we have to learn another way of seeing. As in art, this is an imaginative exercise. Jean-Paul Sartre defined the imagination as the ability to think of what is not present. Thus, the creative imagination must be the chief religious faculty, because it is the only way that we can conceive the eternally absent God. Theology is not about metaphysical facts, which can only be accepted if there is sufficient evidence. It is poetry. Instead of churning out orthodox statements, like my catechism definition of God, theologians like myself should expend as much effort on their God-talk as a poet who is crafting a poem that will touch his readers deeply

FIGURE 8.3. Aristotle and Plato: detail of *School of Athens*, 1510–11 (fresco) by Raphael (Raffaello Sanzio of Urbino) (1483–1520); Vatican Museums and Galleries, Vatican City, Italy.

FIGURE 8.4. Al-Ghazali as portrayed by Ghorban Nadjafi in the film *Al-Ghazali: The Alchemist of Happiness* (2004, directed by Ovidio Salazar). © Matmedia Productions Ltd, www.matmedia.org.

within and lift them momentarily above themselves. Like poetry, theology is a raid on the inarticulate.

As in art, the religious experience of transcendence cannot be achieved by the logic of rational thought. Weighing the evidence of God's existence in a wholly logical way will not give us intimations of the sacred. As all the traditions teach – at their best – the essential prerequisite is that we abandon the egotism and selfishness that hold us back from the divine. They have all devised ways of doing this. Yoga, for example, is an assault on secular consciousness that aims to take the 'I' out of our thinking. The first thing the Prophet Muhammad made his Muslim converts do when he began to preach in Mecca was to prostrate themselves in prayer several times a day. The Arabs did not believe in kingship, and they found it hard to grovel on the ground like a slave, but the posture of their body taught them at a level deeper than the rational how to avoid the preening and prancing ego that is constantly drawing attention to itself. The Buddha promoted the doctrine of *anatta* ('no self'). This was not a metaphysical denial of personality; the Buddha, as we have seen, was not interested in this type of notional truth. His teachings are not about abstractions but describe a methodology. In his view, the ego is the source of much of our distress and suffering; it also makes us harm others. If we lived *as though* the ego did not exist, we would discover the sacred realm of truth in the ground of our being that he called Nirvana. The best and safest way of achieving this transcendence of ego, however, is compassion, which all the world faiths see as the litmus test of true spirituality.

All have taught a version of the Golden Rule, memorably pronounced by Rabbi Hillel, the older contemporary of Jesus. Some pagans had approached Hillel and told him that they would convert to his faith if he could recite the whole of Jewish teaching while he stood on one leg. Hillel replied: 'Do not do unto others as you would not have done unto you. That is the Torah. The rest is commentary. Go and learn it.' Absent here is any mention of God, Mount Sinai or the Promised Land. There is no attempt to convince the pagans by producing evidence to back up the claims of his faith. Hillel simply proposes a course of action that, if practised on a daily, hourly basis, would force us to dethrone ourselves from the centre of our lives and put another there. It brings ecstasy, a word whose literal meaning is 'to stand outside the self'.

Myth and action

Thus, religion is about doing things that in the course of time will transform us and give us new vision. It is not a statement of metaphysical fact but a method. Its truth or effectiveness can only be tested empirically by putting it into practice. Mythology was also a method of action. A myth makes no sense unless it is translated into some form of action. At first this meant ritual action. In the very early societies of humanity, as among indigenous people today, the young men and women usually learned the myths of their tribe in a ritual setting during initiation rites that were designed to transform them into self-reliant adults. A myth only makes sense as part of a process of personal transformation. It tells us how we should confront our demons, how we should enter the labyrinths of our inner selves, how we should face death. However, if it remains at a purely notional level, it remains opaque. It is similar to the instructions of a board game, which sound incomprehensible until we start to play.

Thus the correct approach to a religious myth is not to sift the evidence for its historical truth but to translate it into ritual and ethical action. Take the story of the Israelites' crossing of the Sea of Reeds during the Exodus from Egypt. There have been several well-intentioned efforts, in the spirit of *logos*, to prove the historicity of this tale, by pointing to the prevalence of flash flooding in the region, which could have clogged the chariot wheels of Pharaoh's army. Yet this is to miss the point. If there was a historical event behind this story, it is lost beyond recall. As it has come down to us in the Bible, it has been written up precisely as a myth. In the ancient Near East, gods constantly split a sea in two in the act of creation; what is being brought into being here is not a cosmos but a people. This strange story has become central to the spiritual lives of Jews over the centuries, not because it is founded on a well-documented historical event, but through the Passover Seder. It has been translated from the world of facts and events into the hearts and minds of Jews through ritual action; it has thus become timeless. A myth has been defined as an event that – in some sense – happened once, but which also happens all the time.

Indeed, unless a historical event is mythologized in this way, I doubt that it can become religious. St Paul made the historical Jesus into a *mythos*, not merely by the ritual actions of baptism and the Eucharist, which he describes in his letters, but also through ethical action. This can be seen from one, striking example. In the second chapter of his epistle to the Philippians, Paul quotes what was probably a very early Christian hymn, which described how Jesus had not clung to the fact that he was made in God's image, but had emptied himself, taking the form of a servant; and because of his obedience and *kenosis* (self-emptying) God had raised him to extraordinarily high status and given him the title of Kyrios, or 'Lord'. This is not an early definition of what would later be called the Incarnation. It is a call to action. Paul prefaces the hymn with the exhortation: 'In your minds, *you* must be the same as Christ Jesus'; the myth of Christ's *kenosis* will make no sense unless Christians make it a reality in their own lives, translate it into their daily moral behaviour, and thus prove to themselves that the story is indeed telling us something true about our humanity: that we are most fully ourselves and achieve our full potential, when we give ourselves away.

The misguided quest for certainty

Because of the prevalence of *logos* in our culture, religious people often seem impelled these days to prove how rational their faith truly is. Sometimes this is taken to absurd lengths. I need refer only to the current passion of the Christian Right in the United States for discovering the remains of Noah's ark. Equally absurd is the misnamed 'Creation Science' by which Christian fundamentalists hope to counter the theories of Charles Darwin, finding all kinds of spurious 'evidence' in fossils and the like to back up their claim that the *mythos* of Genesis is true in every detail. Other religious leaders insist that God dictated the whole of the Torah to Moses on Mount Sinai or castigate theologians who suggest that God is not in fact the Supreme Being, as described by classical Western theism. Some try to reconcile themselves with their creeds by pointing out that if these doctrines were entirely credible and obvious, backed up with incontrovertible evidence, there would be no need for the virtue of faith.

I dislike this approach because it seems psychologically and intellectually dangerous to tell yourself that black is white. Moreover, I also think it is essentially unreligious. I am currently writing a book about what might be termed the 'Axial Age' (800 to 200 BCE) in which all the major world faiths emerged at roughly the same time. None of these prophets, sages and mystics had much time for doctrinal conformity. They were not much interested in metaphysics. The Buddha, Socrates and Confucius all insisted that one should question everything, and that it was dangerous to take things at second hand, on faith. The Buddha insisted that his disciples must never take his word for anything, but must only accept his teaching if they found that it worked in their own lives. If not, they should reject it. He liked to tell the story of a traveller who had come to a river, which he desperately needed to cross. But there was no bridge and no ferry, so he cobbled together a raft and paddled himself over. But then, the Buddha would ask his monks, what was he supposed to do with the raft? Should he load it on to his back, because it had been so important to him, and cart it around with him for the rest of his days? Or should he simply moor it, and continue his journey? The answer was obvious. And so, the Buddha would conclude, 'my teachings are like a raft: use them only as long as they help you. If they cease to be of use, put them to one side.' Nothing therefore was cast in stone, and the Buddha, like Jesus I suspect, would be astonished by some modern efforts to prove what are considered to be essential doctrines and beliefs.

But is not faith designed to give the believer certainty? My reading of religious history suggests that it is not. Indeed, Socrates showed that you could not begin your philosophical quest until you had discovered that you could be certain about nothing. The lust for intellectual certainty became particularly acute in the late nineteenth century, when modern discoveries were undermining traditional ideas. People wanted intellectual closure in a world where questions were deliberately left open. It was in this period, for example, that Protestants enunciated the new doctrine of the literal infallibility of Scripture, and Catholics made obligatory the highly controversial doctrine of the infallibility of the Pope.

We have seen too much religious certainty recently. One of the reasons why monotheists insisted that God could not be regarded as a simple personality was that this could easily become an idol. If we consider God to be just like us, but bigger and better, and assume that some of the more dubious remarks attributed to him in scripture are literally true, we are in danger of creating an idol in our own image and likeness. At its very worst, this type of faith can lead to the atrocities of the Crusades or September 11. We can make this 'God' give a sacred seal of absolute approval to our most heinous fears, hatreds and prejudice. But this type of religious certainty can also be 'unskilful' in embedding us in the egotism that we are supposed to transcend. How often we hear preachers maintaining, without a shred of reputable evidence, that God wills this, forbids that and commands the other, and it is uncanny how often the opinions of the deity coincide with those of the speaker. In my view, as I have tried to indicate, the purpose of theology is not to provide us with clear information about God and the supernatural, but to hold us in an attitude of awe and wonder. Religion is at its best when it helps us to ask questions, and at its worst when it seeks to give definitive answers.

FURTHER READING

Armstrong, K., *A History of God*, London: Ballantine, 1993.

Armstrong, K., *Buddha*, London: Phoenix Press, 2002.

Armstrong, K., *The Great Transformation: The World in the Time of Buddha, Socrates, Confucius and Jeremiah*, New York: Atlantic Books, 2006.

Davidson, H. A., *Moses Maimonides: The Man and his Works*, Oxford: Oxford University Press, 2005.

Lewis, B., *Cultures in Conflict: Christians, Muslims and Jews in the Age of Discovery*, Oxford: Oxford University Press, 1996.

McGrath, A., *The Twilight of Atheism: The Rise and Fall of Disbelief in the Modern World*, London: Rider and Co., 2005.

Wolpert, L., *Six Impossible Things Before Breakfast: The Evolutionary Origins of Belief*, New York: Faber and Faber, 2006.

Notes on contributors

Karen Armstrong spent seven years as a Roman Catholic nun in the 1960s, but left her teaching order and went on to study English Literature at the University of Oxford. Since then she has taught Modern Literature at the University of London, headed the English Department in a girls' public school, and taught at the Leo Baeck College for the Study of Judaism and the Training of Rabbis and Teachers. In 1982, she became a full-time writer and broadcaster. Her many books include *Through the Narrow Gate* (1982); *The Gospel According to Woman* (1986); *Holy War: The Crusades and their Impact on Today's World* (1988); *A History of God* (2000); *The Battle for God, A History of Fundamentalism* (2000); *Islam, A Short History* (2004); *The Spiral Staircase* (2005); *A Short History of Myth* (2006); and, most recently, *The Great Transformation: The World in the Time of Buddha, Socrates, Confucius and Jeremiah*. Her books have been translated into 40 languages.

Andrew Bell was educated at University College Oxford. He is Tutor for Admissions and Director of Studies in Anglo-Saxon, Norse and Celtic at Gonville and Caius College, Cambridge, and was previously Moses and Mary Finley Research Fellow at Darwin College, Cambridge, and Lecturer in Medieval History at Magdalen College, Oxford. He is co-author of *From the Dark Ages to the Renaissance 700–1599 AD* (2006).

Cherie Booth QC is a barrister who specializes in Public, Employment and European Community Law at Matrix Chambers, Gray's Inn, London. She is a Recorder and Bencher of Lincoln's Inn, an Honorary Bencher of King's Inn, Dublin, Chancellor and Honorary Fellow of Liverpool John Moores University, and an Honorary Fellow of the London School of Economics and the Open University (D.Univ. Open 1999), LL.D (Hons.) University of Liverpool (2003),

Hon. D.Litt. UMIST (2003). She is also President of Barnardo's, Trustee of Refuge, and Vice President of Kids Club Network.

Vincent Courtillot is a Professor of Geophysics at the Institut de Physique du Globe and University Paris 7. He has taught previously at Stanford, University of California Santa Barbara and Caltech. His research has focused on the Earth's magnetic field, both past and present, and on plate tectonics and Earth geodynamics, about which he has published over 150 papers. His book *Catastrophes in Earth's History: The Science of Mass Extinction* was published by Cambridge University Press in 1999. He is a Fellow of the American Geophysical Union and of the Royal Astronomical Society, and a member of Academia Europaea and the French Academy of Sciences. He is a past President of the European Union of Geosciences and has at various times served as director or special advisor of the French ministries in charge of Education, Research and Technology. He currently chairs the scientific council of the City of Paris.

Philip Dawid studied at Cambridge, where he took a Diploma in Mathematical Statistics and was a graduate member of Darwin College. He is now Professor of Statistics at Cambridge and a Fellow of Darwin. His co-authored book *Probabilistic Networks and Expert Systems* (1999) won the first DeGroot Prize for a Published Book in Statistical Science. As well as publishing widely on theoretical and applied statistics, he has an active interest in the structuring and interpretation of evidence, and currently directs a multidisciplinary research programme on 'Evidence, Inference and Enquiry'.

Carlo Ginzburg, born in Turin in 1939, is currently Franklin D. Murphy Professor of Italian Renaissance Studies at University of California, Los Angeles. His several books include *The Cheese and the Worms* (1980); *The Night Battles* (1983); *The Enigma of Piero* (1985); *Myths, Emblems, Clues* (1990); *Ecstacies: Deciphering the Witches Sabbath* (1991); *History, Rhetoric and Proof* (1999); *No Island is an Island* (2000); *Wooden Eyes* (2001); and (with Pedro Correa do Lago) *True to the Letter: 800 Years of Remarkable Correspondence, Documents and Autographs* (2004). He is an honorary foreign member of the American Academy of Arts and Sciences and a corresponding fellow of the British Academy. He received the Aby Warburg Prize in 1992. His books have been translated into 22 languages.

Monica M. Grady took her first degree at the University of Durham and completed her doctorate in Cambridge as a member of Darwin College. She is currently Professor of Planetary and Space Sciences at the Open University. She was formerly head of the Petrology and Meteoritics Division in the Department of Mineralogy at the Natural History Museum, and Honorary Reader in Geological Sciences at University College, London. She specializes in the study of meteorites, particularly Martian meteorites, and in astrobiology. Her book on astrobiology, *Search for Life*, was published in 2001. Asteroid (4731) was named 'Monicagrady' in her honour. She gave the Royal Institution Christmas Lectures in 2003, on the subject 'A Voyage in Space and Time'.

Brian Greene was educated at Harvard and Oxford, graduating in 1987. Having spent time at Harvard and Cornell, he is currently a Professor of Physics and of Mathematics at Columbia. His recent research work has focused on applying superstring theory to cosmological questions. He has an international reputation as a broadcaster and writer on the fundamental questions of modern science. His best-selling first book, *The Elegant Universe*, was a finalist for the Pulitzer Prize in General Nonfiction and won the Aventis Prize in 2000. His most recent book, *The Fabric of the Cosmos* (2004), has received much critical acclaim, spending ten weeks on the *New York Times* best-seller list.

Peter Lipton was the Hans Rausing Professor and Head of the Department of the History and Philosophy of Science at Cambridge University, and a Fellow of King's College. His main philosophical interests lay in the theory of knowledge and the philosophy of science, especially questions concerning explanation and inference. These are questions about the difference between knowing that a given phenomenon occurs and understanding why it occurs, and questions about the ways the reliability of inconclusive evidence is judged. The second edition of his book *Inference to the Best Explanation* was published in 2004. Peter Lipton died in 2006.

John Swenson-Wright is University Lecturer in Modern Japanese Studies at the East Asia Institute, Faculty of Oriental Studies, University of Cambridge, and an official Fellow of Darwin College. His research and teaching focus is on the international relations and politics of East Asia, with particular reference to Japan and the Korean peninsula. He is the author of *Unequal Allies? United*

States Security and Alliance Policy Towards Japan, 1945–1960 (2005) and is the editor of a translated memoir by Wakaizumi Kei, *The Best Course Available: A Personal Account of the Secret US–Japan Okinawa Reversion Negotiations* (2002).

Karin Tybjerg is Leader of the Department of Astronomy at Kroppedal Museum of Astronomy, Archaeology and Modern History in Copenhagen and an Associated Scholar of the Department of History and Philosophy of Science at the University of Cambridge. She was educated in Physics, Philosophy and History of Science at King's College London and the University of Cambridge, and held the Moses and Mary Finley Fellowship at Darwin College, Cambridge. Her research interests focus on ancient science and mathematics and the history of astronomy.

Index

Printed in the United States
By Bookmasters